Carbon Nanostructures

Caro Stephen,
con stima
e gratitudine

A presto

[illegible]

More information about this series at http://www.springer.com/series/8633

Vittorio Morandi · Luca Ottaviano
Editors

GraphITA

Selected papers from the Workshop on Synthesis, Characterization and Technological Exploitation of Graphene and 2D Materials Beyond Graphene

Springer

Editors
Vittorio Morandi
CNR-IMM Section of Bologna
Bologna
Italy

Luca Ottaviano
Università degli Studi dell'Aquila
L'Aquila
Italy

ISSN 2191-3005 ISSN 2191-3013 (electronic)
Carbon Nanostructures
ISBN 978-3-319-58132-3 ISBN 978-3-319-58134-7 (eBook)
DOI 10.1007/978-3-319-58134-7

Library of Congress Control Number: 2017940309

Printed on acid-free paper

This Springer imprint is published by Springer Nature
The registered company is Springer International Publishing AG
The registered company address is: Gewerbestrasse 11, 6330 Cham, Switzerland

Preface

This volume contains selected invited papers presented at GraphITA (Bologna Italy September 14–18, 2015—http://graphita.bo.imm.cnr.it), a multidisciplinary and inter-sectorial European conference on synthesis, characterization and applications of Graphene and other 2D materials. The event was jointly organized by the CNR-IMM (Consiglio Nazionale delle Ricerche, Istituto per la Microelettronica e Microsistemi) of Bologna and the Department of Physical and Chemical Sciences of the University of L'Aquila. At variance from GraphITA 2011, this volume is focused on invited review papers mostly contributed by the invited speakers of the conference. Though, all the contributors are, in their respective fields, top players internationally acknowledged, however all the submitted papers were submitted to a peer review process.

This volume begins with closely related and complementary theoretical works on phonon, Raman scattering, and electron phonon coupling in graphene (Chapters "Thermal Transport in Nanocrystalline Graphene: The Role of Grain Boundaries", "Raman Spectroscopy of Graphene Nanoribbons: A Review", and "Electron–Phonon Coupling in Two-Dimensional Superconductors: Doped Graphene and phosphorene" respectively). The latter point is then reviewed with an excellent experimental contribution (Chapter "Elastic Properties and Electron–Phonon Coupling of Graphene/Metal Interfaces Probed by Phonon Dispersion"). Then, the volume proceeds with an ab initio study of the doping effect on graphene, to move towards the investigation of the role of defects on other 2D materials (TMDs) (Chapters "*Ab Initio* Calculations and Kinetic Process Simulations of Nitrogen-Doped Graphene", and "From Point to Line Defects in Two-Dimensional Transition Metal Dichalcogenides: Insights from Transmission Electron Microscopy and First-Principles Calculations" respectively). The paper presented in Chapter "Morphing Graphene-Based Systems for Applications: Perspectives from Simulations" links the first part of the volume with the second and addresses the issue of tailoring the modelling approaches in the simulations of graphene-based systems towards a more realistic comparison with experiments at coarse scales. In Chapter "Perfecting the Growth and Transfer of Large Single-Crystal CVD Graphene: A Platform Material for Optoelectronic Applications", an up-to-date survey of the real world of CVD growth of graphene is presented. This chapter paths the way for the rest of the volume which is more focused on applications of graphene and other 2D systems: electronics (in Chapter "Advances in the Fabrication of Large-Area Back-Gated Graphene Field-Effect Transistors on Plastics: Platform for Flexible Electronics and Sensing"), growth of other 2D "enes" (Chapter "Silicene in the Flatland"), energy storage (Chapter "Decorated and Modified Graphenes as Electrodes in Na and Li-Ion Batteries"), gas sensing (Chapter "Chemically Exfoliated Layered Materials for Practical Gas Sensing Applications"), functional materials generation (Chapter "Solutions of Reduced Carbon Allotropes and Their Utilization for Functional Material Generation"), fabrication of foams (Chapter "Synthesis of High-Density Graphene Foams using Nanoparticle Templates") and novel advanced functional materials based on proteins and graphene (Chapter "Protein-Based Nanostructures and Their Self-assembly with Graphene Oxide"), to finish with a survey of nanobio applications of graphene biocomposites (Chapter "Graphene Bionicomposites").

In summary, although completeness is by far too an ambitious goal in a single-volume survey of the exploding multidisciplinary research on the world of 2D materials, this book is however a *nice journey through 2D materials from fundamentals and applications.*

As Editors, we are very grateful to all the members of the International Advisory Committee, as well as other anonymous referees, for their valuable contribution to the review procedure. Moreover, as Chairs, we would like to thank all the members of the International Scientific Committee and of the Local Committee of GraphITA for their fundamental contribution and support for the success of the workshop. Finally, we are very grateful to Mayra Castro, and Petra Jantzen of Springer Office for their helpful assistance during the preparation of this special volume.

Bologna, Italy
L'Aquila, Italy
2015

Vittorio Morandi
Luca Ottaviano
The Editors and Chairs of GraphITA

Contents

Thermal Transport in Nanocrystalline Graphene: The Role of Grain Boundaries

K.R. Hahn, C. Melis and L. Colombo

Abstract Single grain boundaries of crystalline graphene with varying mismatch angles from 3° to 16° have been investigated using molecular dynamics simulations. Four- to eight-atomic rings are found to be the most abundant non-hexagonal polygons in the grain boundary for all mismatch angles. Tetra- and octagons are predominant for mismatch angles of 4.1° and 6.6° in contrast to nanocrystalline samples where penta- and heptagons are dominating. Out-of-plane buckling at the grain boundary is most pronounced for a mismatch angle of 3.0° and it tends to decrease with increasing mismatch angle. At 16.1°, the out-of-plane buckling vanishes. Analysis of the vibrational density of states of boundary atoms revealed a significant decrease of the main peak of optical vibrations and the evolution of secondary peaks below and above the major frequency attributed to vibrations of non-hexagonal rings. The thermal boundary resistance in single graphene interfaces has been approximated. It tends to increase with increasing mismatch angle, indicating reduced thermal conductivity when such interfaces are present in crystalline graphene. In nanocrystalline graphene samples, the thermal conductivity is significantly reduced with respect to crystalline graphene and it decreases with decreasing grain size according to an increasing number of single boundaries.

K.R. Hahn · C. Melis · L. Colombo (✉)
Department of Physics, University of Cagliari, Cittadella Universitaria, 09042 Monserrato, CA, Italy
e-mail: luciano.colombo@unica.it

K.R. Hahn
e-mail: konstanze.hahn@dsf.unica.it

C. Melis
e-mail: claudio.melis@dsf.unica.it

L. Colombo
Institut de Ciència de Materials de Barcelona (ICMAB-CSIC), Campus UAB, 08193 Bellaterra, Barcelona, Spain

L. Colombo
Catalan Institute of Nanoscience and Nanotechnology (ICN2), CSIC and the Barcelona Institute of Science and Technology, Campus UAB, 08193 Bellaterra, Barcelona, Spain

V. Morandi and L. Ottaviano (eds.), *GraphITA*, Carbon Nanostructures,
DOI 10.1007/978-3-319-58134-7_1

1 Introduction

Monolayer graphene is typically fabricated by epitaxial film growth [1, 2] and chemical vapor deposition (CVD) [3–5], where the latter is promising in particular for large-scale production [3, 6]. A common aspect of all fabrication methods is that defects are generated in the honeycomb lattice resulting from limitations of the kinetic properties in the growth process and defects of the substrates. These defects limit the size of single-crystalline domains and graphene sheets should therefore be looked at as polycrystalline. Grain boundaries, however, drastically change the mechanical and electronic properties of the system [7–9]. The polycrystalline nature with dimensions down to the micro- and nanoscale is also supposed to be responsible for the wide range of values of the thermal conductivity measured experimentally in graphene (600–10000 W/mK [10–13]).

Several experimental, theoretical, and combined studies have been performed on the investigation of grain boundaries in graphene [14–22]. The systematic investigation of the effect of size, shape, and distribution of grains especially on the thermal conductivity of graphene is still ongoing. In particular, the thermal conductivity of nanocrystalline graphene generated by grain growth, has been estimated using approach-to-equilibrium molecular dynamics (AEMD) simulations proving a strong dependence of the thermal conductivity on the grain size [23]. Other studies have applied equilibrium molecular dynamics (MD) simulations and the Green–Kubo expression for the investigation of the thermal conductivity in nanocrystalline graphene [24, 25].

Here we investigate the effect of the mismatch angle in single interfaces of tilted graphene on its morphology, the vibrational density of states (VDOS), and the thermal boundary resistance. Furthermore, we have generated nanocrystalline graphene with grain sizes up to 20 nm using two different methods. In the first, nanocrystalline graphene has been generated where grains have been grown by an iterative algorithm (bottom-up) [23]. The other one uses the Voronoi tessellation scheme to generate a pattern of crystalline domains which have been cut-out of a single-crystalline sample cell (top-down) [26]. The structure of the generated simulation domains has been determined and their thermal conductivity has been investigated applying AEMD simulations [27].

2 Methods

2.1 Sample Generation

2.1.1 Single Interfaces

Single grain boundaries in graphene have been created by tilting two single-crystalline graphene sheets of the same length by a certain angle one versus the other. To match periodic boundary conditions, only some specific orientations are possible as described previously [19, 28]. Here, tilt angles α of 3.0°, 4.1°, 8.2°, 10.9°, and 16.1° have been studied. The definition of the tilt angle is shown in Fig. 1 on the example of a grain boundary with α = 8.2°.

2.1.2 Nanocrystalline Graphene by Grain Growth Method

In analogy to experiments, generation of nanostructured systems for theoretical investigations can be realized either by bottom-up or top-down methods. In top-down methods, first, homogeneous amorphous or crystalline simulation cells are created which are then modified by scissoring and rearrangement of partially homogeneous sections to create nanostructured materials. In contrast, in bottom-up methods, nanostructures are constructed from single atoms or small building blocks (for example, crystalline seeds) and simulation techniques such as MD are used to grow the nanostructures from the initial seeds. Here, simulation cells have been generated using both methods. The grain growth method is a bottom-up approach while the Voronoi method can be regarded as a top-down approach.

For the grain growth method, an algorithm has been developed where, for a given rectangular simulation cell with width L_x and length L_z, a specified number of nucleation sites has been placed at random positions in the cell with the constraint that neighboring seed atoms have a certain distance between each other. The number of nucleation sites determines the dimension of the generated grains. Here, densities from 0.32 to 3.7 × 10^{-4} seeds/nm^2 have been used to obtain grain sizes from 0.7 to 20 nm.

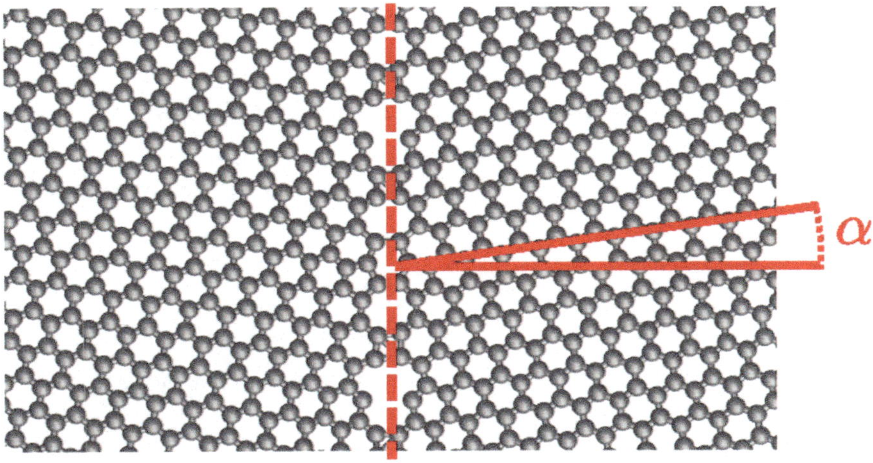

Fig. 1 Graphene interface with a mismatch angle α of 8.2°

An arbitrary angle between 0 and 60° has been assigned to each nucleation site to define the orientation of the crystalline growth of graphene. Under-coordinated atoms on the edge of each grain have been added to a list of possible reactive sites. In an iterative process, the reactive site has been randomly selected from this list to which the next atom has been added. The list of reactive atoms has been updated in each step until no reactive atoms were left. Grain growth has been terminated, i.e., atoms have been eliminated from the list of reactive atoms, when the respective C atom was threefold coordinated or atoms from neighboring cells were closer than 95% of the equilibrium C–C bond length (here: 1.3978 Å). The method has been described in more detail previously [23]. A section of a sample cell is shown in Fig. 3a.

2.1.3 Nanocrystalline Graphene by Voronoi Tessellation

First the cell dimensions and randomly chosen seed sites have been defined. The simulations cells have been divided into regions based on the seed positions using the Voronoi tessellation method. The Voronoi formalism defines a region R_p associated with the seed s_p according to

$$R_p = \left\{x \in R^2 | \delta(x, s_p) \leq \delta(x, s_j) \text{ for all } j \neq p\right\}, \tag{1}$$

where x is the position in the simulation cell, s_j the center of the other seeds and δ is the Euclidean distance. Each of the so-defined regions has been cut-out from a template cell of crystalline graphene and rotated by a randomly assigned angle. A section of a sample cell is shown in Fig. 3b. At least two different configurations have been created for each sample dimension and grain size. The average radius of gyration (definition see below) has been changed from 1 to 20 nm using samples with a cell length of 200 nm.

2.2 *Structural Characterization of Generated Samples*

The grain size in nanocrystalline graphene has been characterized by the radius of gyration r_{G} according to

$$r_{\mathrm{G},j}^2 = \frac{\sum_i^{n_{\mathrm{C},j}} \left(\mathbf{r}_{i,j} - \mathbf{r}_{CM,j}\right)^2}{n_{\mathrm{C},j}}, \tag{2}$$

where $\mathbf{r}_{i,j}$ is the position of atom i in grain j and $\mathbf{r}_{CM,j}$ and $n_{\mathrm{C},j}$ are the center of mass and the number of atoms in the grain, respectively.

Grain boundary atoms both in single interfaces and in nanocrystalline graphene have been identified by their structural properties. All C atoms that do not belong to a hexagonal ring have been selected by this method. The distribution of polygonal rings in the entire simulation cell has been determined by iteratively following the nearest and next-nearest neighbor positions of each atom. To characterize the grain boundaries, the number of non-hexagonal polygons at the grain boundary has been calculated and their distribution has been determined.

Strain at the interface leads to a deformation of the graphene sheet perpendicular to the sheet plane. This corrugation is denoted here as out-of-plane buckling. In single interfaces, it is characterized assuming a Gaussian-type distribution for the out-of-plane coordinates $y(z)$ of C atoms with the maximum at the interface according to

$$y(z) = A \cdot e^{\frac{(z-\mu)^2}{\sigma^2}}. \tag{3}$$

Using this, the spread of the out-of-plane buckling is determined by the deviation σ and the pre-factor A which corresponds to the amplitude of the buckling.

In nanocrystalline graphene, on the other hand, analysis directly at the interface is not possible. The degree of out-of-plane buckling is therefore analyzed by the average position of all atoms in the simulation cell. From the arithmetic mean and the standard deviation, a standard normal distribution can be extrapolated [26].

In some simulation cells, a sinusoidal deformation of the graphene sheet has been observed which is periodic in the cell length L_z. To eliminate the effects of this long-period deformation, the dislocation of C atoms has been calculated with respect to this sinusoidal function which has been fitted to the out-of-plane position of all atoms in the simulation cell [26].

2.3 AEMD Simulations

Generation of samples as described above only considers geometrical aspects. Interatomic interactions are neglected and the samples are created only in two dimensions, omitting corrugation in out-of-plane direction. To account for the latter, simulated annealing has been applied to all generated structures. It has been initiated at a temperature of 300 K for 10 ps. Subsequently, the polycrystalline samples (samples with single boundaries) have been heated to a temperature of 4000 K (1000 K) and relaxed at this temperature for 300 ps (50 ps). Then samples have been cooled back to 300 K and further aged.

Molecular dynamics simulations have been performed using the LAMMPS code [29, 30] and covalent interactions between carbon atoms have been described by the second-generation reactive empirical bond order (REBO) potential which has previously been shown to give reasonable results for graphene-based systems [23, 31, 32]. The thermal conductivity has been determined based on the AEMD methodology. Details are reported elsewhere [27].

3 Results and Discussion

3.1 Structural Properties

3.1.1 Polygon Distribution

At the grain boundary of crystalline graphene domains, carbon rings with different number of atoms are formed. The distribution of the carbon polygons is shown in Fig. 2. It has been normalized to the number of non-hexagonal rings. In all samples, non-hexagonal rings mostly consist of four to eight atoms in addition to some ten-sided polygons. In grain boundaries with a mismatch angle of 6.6° and 16.1° the number of ten-sided polygons can reach up to 6%.

At a mismatch angle of 3.0°, the number of polygons with four to eight atoms is almost equally distributed with a slight tendency towards tetragons and octagons. This tendency becomes more pronounced with increasing mismatch angle to 4.1° and 6.6° resulting in a remarkable decrease in pentagons and heptagons in particular at $\alpha = 6.6°$. The high population of tetragons and octagons at 4.1° and 6.6° is in agreement with previous results where two peaks in the ring population are found at chiral angles close to 30° [33] which correspond to small mismatch angles as defined in our study. Similar numbers for all polygons are found at a mismatch angle of $\alpha = 10.9°$ with a tendency towards octagons. In grain boundaries with $\alpha = 16.1°$, the highest population of pentagon rings with ca. 30% is observed.

In nanocrystalline samples, heptagon and pentagon rings are predominant indicating that mismatch angles in polycrystalline graphene are in the range of 8°–16° (Fig. 3c) [26]. However, it has to be noted that grain boundaries in nanocrystalline graphene intrinsically do not form straight lines as the ones simulated in the single grain boundaries. The nature of grain boundaries in

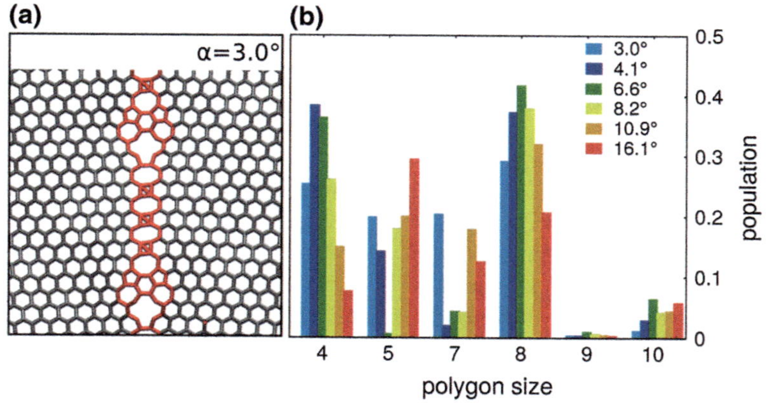

Fig. 2 **a** Grain boundary of tilted graphene with a mismatch angle of 3.0°. **b** Distribution of polygon rings at the grain boundary

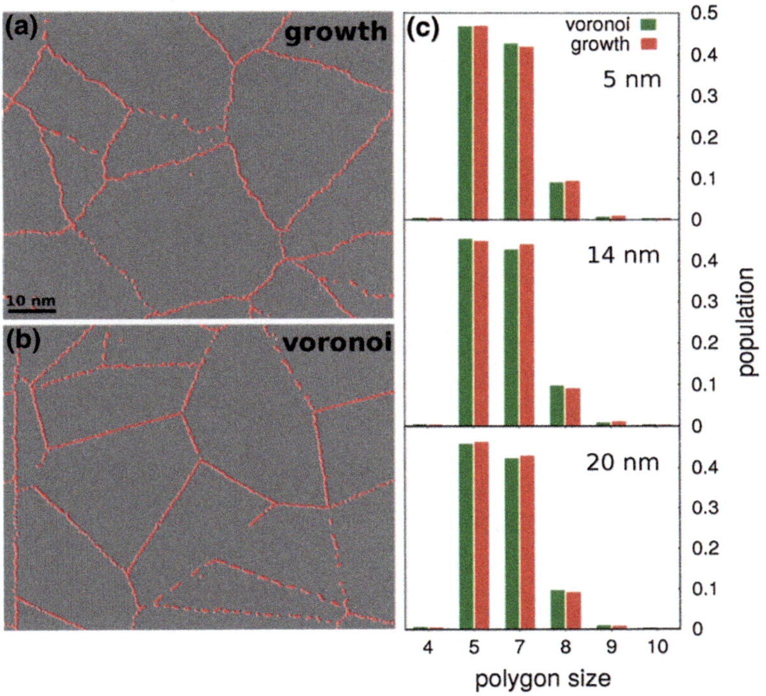

Fig. 3 Samples of nanocrystalline graphene generated by **a** the grain growth and **b** the Voronoi tessellation method. **c** Distribution of polygon rings in nanocrystalline graphene with average grain size of $r_G = 5$, 14 and 20 nm generate by the Voronoi (*green*) and the grain growth (*red*) method

nanocrystalline graphene is directed by the energetically most convenient structure resulting in boundaries that are not necessarily straight (Fig. 3a, b) and thermodynamically favorable ring structures are formed.

Grain boundaries in the Voronoi samples form more direct lines compared to grain growth samples. In fact, a tendency towards octagon formation is noted in nanocrystalline samples with larger grain size (>14 nm) [26]. This is in agreement with previous experiments where straight grain boundaries have been generated experimentally in graphene grown on Ni(111) [34]. Such grain boundaries have been found to consist mostly of a combination of pentagonal and octagonal C-rings.

3.1.2 Out-of-Plane Buckling

The generated strain at the grain boundary leads to an out-of-plane corrugation of the graphene sheet (Fig. 4). The magnitude of such displacement of C atoms perpendicular to the graphene plane as a function of the distance from the grain

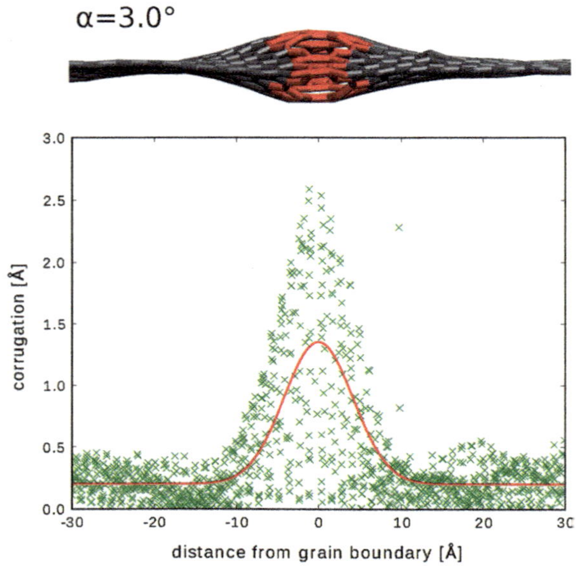

Fig. 4 Distribution of out-of-plane coordinate of C atoms at the grain boundary for an interface with atomic mismatch of $\alpha = 3.0°$

boundary can be characterized by a Gaussian distribution as described in the methods section (Eq. 3).

The out-of-plane buckling in the grain boundaries of all samples has been fitted to the Gaussian distribution and the obtained values for the spread σ and the height (magnitude) A have been averaged over all obtained values for one mismatch angle. Since atomic positions of all atoms have been used to fit the parameters σ and A, the obtained height A is an average value and thus significantly smaller than the largest deviation from the graphene plane. As shown in Fig. 4, the buckling directly at the grain boundary is about 2.5 Å (whereas the approximated height of the Gaussian function is 1.4 Å). This is in agreement with previous studies where out-of-plane buckling has been reported to be between 1.5 and 2.5 Å [19]. The values for the fitted magnitude A of the buckling is between 0.84 and 1.28 Å for mismatch angles from 3° to 10.9° and it tends to decrease with increasing mismatch angle (Fig. 5).

The spread σ of the interface buckling, on the other hand, does not show a trend with the mismatch angle. It is largest for a mismatch angle of 3.0° and 10.9° and slightly lower for mismatch angles between 4.1° and 8.2°. This can possibly be correlated to the larger number of tetra- and octagons in grain boundaries with angles between 4.1° and 8.2°. The out-of-plane buckling in grain boundaries with a mismatch angle of 16.1° is significantly different to the buckling with smaller mismatch angles. Here, basically no buckling is observed and both magnitude A and spread σ are close to zero. This is in agreement with previous results, in particular the vanishing out-of-plane buckling at $\alpha = 16.1°$ corresponding to $\theta = 32.2°$ of the previous study [35].

In nanocrystalline graphene, out-of-plane buckling has been characterized considering the out-of-plane displacement of all C atoms in the sample. Samples

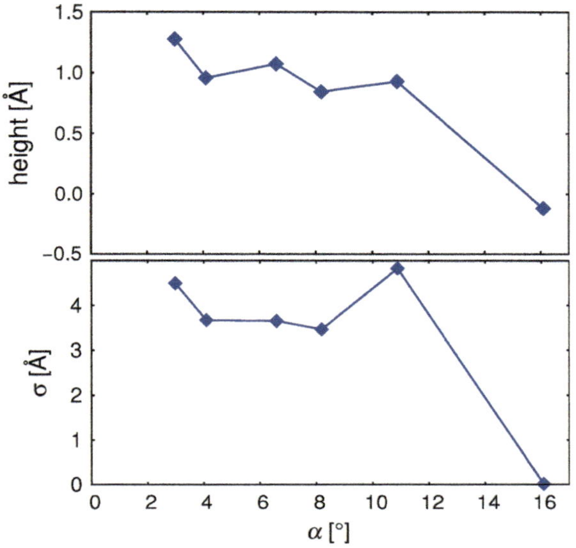

Fig. 5 Magnitude (*height*) and spread (σ) of the out-of-plane buckling at the interface as a function of the mismatch angle

generated by the Voronoi method show larger out-of-plane buckling with respect to grain growth samples, especially for grain sizes smaller than 15 nm [26].

3.1.3 Vibrational Density of States

The vibrational density of states (VDOS) has been calculated as the Fourier transform of the velocity autocorrelation function collected during a canonical simulation of 300 K for 10 ps. For each mismatch angle, the smallest simulation cell ($L_z \approx 100$ nm) has been used to analyze the VDOS. Velocities have been considered for atoms within 10 Å of the grain boundary. Figure 6 shows the VDOS of these grain boundary atoms in comparison to the VDOS of single-crystalline graphene. The VDOS of crystalline graphene is in good agreement with previous calculations [36]. Three main peaks can be distinguished in the acoustic vibrations of single-crystalline graphene which are found at 12, 21, and 30 THz. In the grain boundary, the main peak at 21 THz is still apparent and most pronounced for a mismatch of $\alpha = 6.6°$. For mismatch angles of 3.0°, 10.9°, and 16.1° (Fig. 6), the valley between the 12 and 21 THz peak becomes less pronounced which results from additional vibrations in this frequency range. This can be a result of increased concentration of defects in this range.

One main peak at 53 THz exists in the optical phonon spectrum of graphene and is attributed to the so-called *G* peak resulting from bond stretching of sp^2 atoms in rings and chains [37]. Its position can change when the 2D graphene sheets are exposed to strain [38, 39]. In all samples of single grain boundaries in tilted graphene, the intensity of the *G* peak is significantly reduced with respect to defect-free

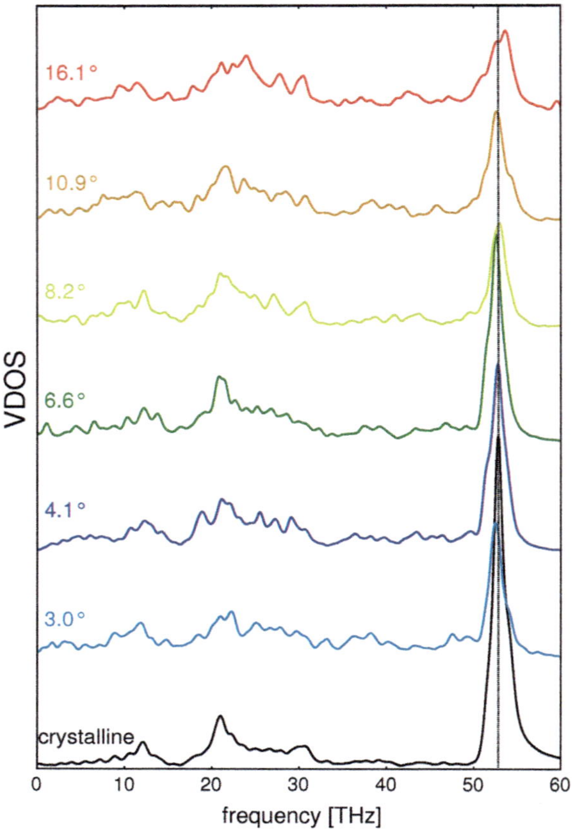

Fig. 6 Vibrational density of states of the grain boundary atoms in graphene interface with different mismatch angles

crystalline graphene, in particular for mismatch angles of 3.0°, 8.2°, 10.9°, and 16.1° (Fig. 6).

Marginal deviations of the *G* peak position are found for mismatch angles of 4.1° and 6.6°. At both interfaces a small shoulder evolves at slightly lower frequencies, resulting in a broadening of the peak. At a mismatch angle of 3.0°, 8.2°, and 10.9°, the *G* peak is red-shifted to a frequency f_{Gmax} of ca. 52.5 THz. In addition, the intensity of the peak is reduced and a shoulder at higher frequencies appears in the sample with $\alpha = 3.0°$ and at both higher and lower frequencies in the sample with $\alpha = 10.9°$.

Most remarkable differences in the *G* peak appear in grain boundaries with a mismatch of 16.1°. There, the peak maximum is shifted to a higher frequency ($f_{Gmax} = 53.6$ THz) with a significantly lower intensity with respect to the peak maximum in single-crystalline graphene. Furthermore, the peak is broadened resulting from additional vibrations at lower frequencies. In general, the evolution of additional peaks in all samples can be attributed to vibrations in non-hexagonal polygons. It is suggested that the higher frequency in the 16.1° sample results from stretching vibrations of C–C bonds belonging to non-ideal hexagon rings (like, e.g., penta- and heptagons and/or distorted C–C bonds lying at the grain boundary). With respect to the other samples, the number of pentagons in the 16.1° grain boundary is highest (see Fig. 2).

A red-shift of the *G* peak is also observed in nanocrystalline graphene samples accompanied by the evolution of two minor peaks at higher and lower frequencies [26]. The additional peaks are more pronounced with respect to the spectrum of single grain boundaries and can be attributed to vibrations in pentagon and heptagon rings which are the prominent non-hexagonal structures in grain boundaries of nanocrystalline graphene (Fig. 3c).

3.2 Thermal Transport Properties

3.2.1 Thermal Boundary Resistance in Single Boundaries

Generally, the total thermal resistance R_{tot} in a material with an interface can be described as a connection of series of resistances according to

$$R_{tot} = \frac{L_z}{\kappa_{tot}} = \frac{L_a}{\kappa_a} + R_{if} + \frac{L_b}{\kappa_b}, \tag{4}$$

where κ_{tot} is the overall thermal conductivity of a system with length L_z along the direction of the thermal flux, κ_a and κ_b are the thermal conductivities in the homogeneous sections *a* and *b*, respectively, which have the respective length L_a and L_b. R_{if} is the thermal resistance of the interface.

In the graphene interfaces simulated here, sections *a* and *b* correspond to crystalline graphene where $L_a = L_b = L_z/2$ and the thermal conductivities $\kappa_a = \kappa_b$ correspond to the one of single-crystalline graphene (κ_{sc}). Considering two interfaces in the simulation cell, Eq. (4) can be written as

$$\begin{aligned} \frac{L_z}{\kappa_{tot}} &= \frac{L_z}{\kappa_{sc}} + 2R_{if} \\ R_{if} &= \frac{L_z}{2}\left(\frac{1}{\kappa_{tot}} - \frac{1}{\kappa_{sc}}\right) \end{aligned} \tag{5}$$

In simulations with finite cell sizes, the thermal conductivity is a function of the cell length and is commonly described by a linear relationship of the inverse values $\frac{1}{\kappa} = \frac{1}{\kappa_\infty}\left(1 + \frac{\lambda}{L_z}\right)$ [40–42], where κ_∞ is the bulk thermal conductivity at an infinite cell length $L_z \rightarrow \infty$ and λ the characteristic length of the phonon spectrum. Using this approximation and the assumption that $\frac{1}{\kappa_{tot,\infty}} \approx \frac{1}{\kappa_{sc,\infty}}$ (at infinite cell length), the thermal boundary resistance R_{if} can be calculated by

$$R_{if} = \frac{1}{2}\left(\frac{\lambda_{tot}}{\kappa_{tot,\infty}} - \frac{2\lambda_{sc}}{\kappa_{sc}}\right). \qquad (6)$$

Based on this, R_{if} is evaluated as a function of the mismatch angle α (Fig. 7). The smallest thermal boundary resistance of 0.14 m^2K/GW is found for a mismatch of $\alpha = 3.0°$ and it increases to 0.38 m^2K/GW with increasing α up to 6.6°. A similar thermal boundary resistance of ca. 0.3 m^2K/GW results for $\alpha = 4.1°$, 8.2°, and 10.9°. The highest R_{if} of 0.58 m^2K/GW is found in samples with largest atomic mismatch of $\alpha = 16.1°$. Similar values of the thermal boundary resistance have been measured experimentally for a mismatch angle of 3.0° (0.44 m^2K/GW) [43]. These experiments furthermore confirm the general trend of increasing thermal boundary resistance with increasing mismatch angle.

Previously, enhanced phonon scattering and thus increased thermal boundary resistance have been attributed to larger out-of-plane buckling [26, 35]. This, however, is not valid for the systems calculated here. The highest thermal boundary resistance (Fig. 7) occurs in grain boundaries with the largest mismatch angle 16.1° where the lowest out-of-plane buckling is observed (Fig. 5).

3.2.2 Thermal Conductivity in Nanocrystalline Graphene

Grain size effects on the thermal conductivity in nanocrystalline graphene have been investigated in simulation cells with a sample length $L_z = 200$ nm. In the interest of computational cost and time, the estimation of the thermal conductivity by AEMD simulations has been limited to only one specific cell length and the resulting thermal conductivity is referred to the thermal conductivity of single-crystalline graphene at the identical sample length $L_z = 200$ nm (262 W/mK).

Figure 8 shows the relative thermal conductivity of nanocrystalline graphene generated by grain growth and the Voronoi method as a function of the grain size

Fig. 7 Thermal boundary resistance R_{if} as a function of the mismatch angle α

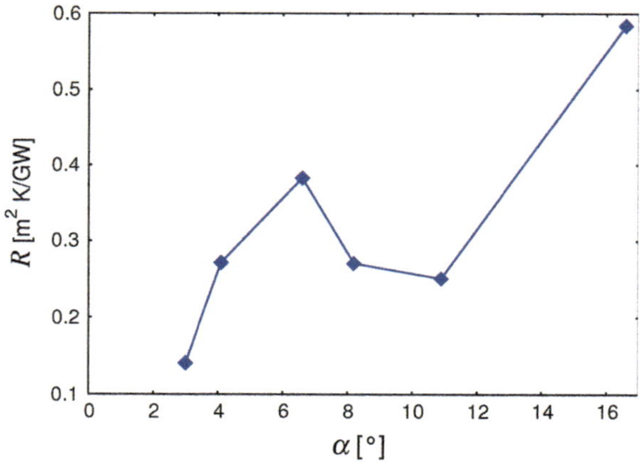

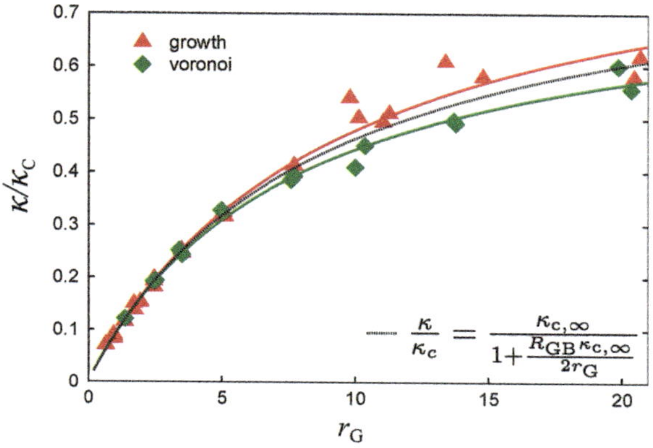

Fig. 8 Relative thermal conductivity (κ/κ_c) normalized to the thermal conductivity of single crystalline graphene ($\kappa_c = 262$ W/mK) as a function of the radius of gyration r_G. Results of the fitting parameters are shown in Table 1

Table 1 Parameters fitted to Eq. (7) for the thermal conductivity in nanocrystalline graphene as a function of r_G

Generation method	$\kappa_{c,\infty}$[–]	$\frac{R_{GB}\kappa_{c,\infty}}{2}$ [nm]
Growth	0.925	9.29
Voronoi	0.783	7.67
Both	0.858	8.52

(r_G). It increases continuously with increasing grain size for both methods. Despite the identical general trend, small differences in the thermal conductivity are observed for grain sizes between 5 and 15 nm. In this region, nanocrystalline graphene with grains generated by the grain growth method shows a higher thermal conductivity than the Voronoi samples.

A possible explanation for this difference is the higher degree of corrugation that is observed for grain sizes ≤ 14 nm in samples generated by the Voronoi method [26]. Previously, MD simulations have evidenced enhanced scattering of long-wavelength flexural phonons at grain boundaries with larger out-of-plane buckling [35]. The out-of-plane corrugation is observed to be even larger at smaller grain sizes in the Voronoi-produced samples [26]. In nanocrystalline materials, however, the phonon spectrum is reduced to the dimension of the crystalline domains, and it is expected that the dimension of the crystalline domains for small grain sizes (≤ 5 nm) is the limiting factor for the thermal transport. Therefore, no difference in the thermal conductivity is observed between the two generation methods for small grains. The effect of increased phonon scattering of long-wavelength phonons at corrugated grain boundaries is only observed at larger grain sizes (≥ 5 nm).

Grain boundaries can be regarded as a connection in series of resistances [23, 24]. This argumentation leads to a rational function to describe the κ versus r_G behavior according to

$$\frac{\kappa}{\kappa_c}(r_G) = \frac{\kappa_{c,\infty}}{1 + \frac{R_{GB}\kappa_{c,\infty}}{2r_G}}. \tag{7}$$

The parameters $\kappa_{c,\infty}$ and R_{GB} have been optimized separately for the two generation methods. As a result of the lower thermal conductivity for grain sizes between 5 and 15 nm in the Voronoi samples, the relative ideal thermal conductivity in crystalline graphene ($\kappa_{c,\infty}$) has been determined to be lower (0.78) than for cells generated by grain growth (0.92). With both methods and the assumptions of Eq. (7), the relative thermal conductivity should converge to 1 for infinitely large grains. For grown grains, the optimized value (0.92) is reasonably close.

In Eq. (7), a stepwise uniform thermal conductivity is assumed. With the enhanced scattering of flexural phonons at corrugated grain boundaries, however, additional scattering effects are present which might lead to a nonlinear behavior of the thermal conductivity. Therefore a lower value of $\kappa_{c,\infty}$ is found for the Voronoi-patterned graphene cells.

The rough estimation of the thermal boundary resistance by the optimized parameter R_{GB} resulted in 7.7×10^{-11} m^{2}K/W and 8.1×10^{-11} m^{2}K/W for the grain growth and the Voronoi method, respectively. This result suggests that the topology of grains only marginally affects thermal resistance properties.

4 Conclusions

The effect of the mismatch angle in tilted graphene on morphology, VDOS and thermal conductivity has been investigated using atomistic simulations. The distribution of polygonal rings at the interface is significantly affected by the mismatch angle. Non-hexagonal rings in the grain boundaries consist mostly of four to eight atoms, with a prevalence of tetra- and octagons for mismatch angles of 4.1° and 6.6°. In nanocrystalline graphene, grain boundaries intrinsically are not straight allowing for the formation of thermodynamically most favorable ring structures. In fact, in the samples generated here by two different methods, pentagon and heptagon rings are predominant.

Strain created at the interface leads to a corrugation of C atoms perpendicular to the graphene plane. Similar out-of-plane buckling is found for mismatch angles from 3.0 to 10.9° where the magnitude of corrugation tends to decrease with increasing mismatch angle. Strikingly, at a mismatch angle of 16.1°, the out-of-plane buckling basically vanishes in agreement with previous studies.

Acoustic vibrations of the grain boundary atoms vary marginally from the vibrational spectrum of single-crystalline graphene. Differences, however, are observed in the position and intensity of the main peak (*G* peak) of the optical phonons. The intensity of the peak is reduced with respect to single-crystalline graphene as a result of the formation of non-hexagonal C-rings. The *G* peak of vibrations of grain boundaries atoms in nanocrystalline graphene shifts to lower frequencies with respect to the spectrum in single crystalline graphene. This is more pronounced in cells generated by the grain growth method and it becomes stronger with decreasing grain size. It can be attributed to tensile strain which is more pronounced in grain growth samples with small grain sizes ($r_G < 5$ nm).

The estimated thermal boundary resistance in single boundaries of tilted graphene tends to increase with increasing mismatch angle. The highest value is observed for a mismatch angle of 16.1° with 0.55 m^2K/GW. Despite significant differences in the shape of the grains and the particle size distribution in nanocrystalline graphene, the thermal conductivity in samples with the same average radius of gyration is nearly the same in samples generated with different methods. For a grain size between 5 and 15 nm, simulation cells generated by the Voronoi method have a slightly lower thermal conductivity than the grown samples. This could possibly result from the larger out-of-plane corrugation in the Voronoi-generated samples for small grain sizes (<15 nm).

References

1. Berger, C., Song, Z., Li, X., Wu, X., Brown, N., Naud, C., Mayou, D., Li, T., Hass, J., Marchenkov, A.N., Conrad, E.H., First, P.N., de Heer, W.A.: Electronic confinement and coherence in patterned epitaxial graphene. Science **312**, 1191–1196 (2006)
2. Hass, J., de Heer, W.A., Conrad, E.H.: The growth and morphology of epitaxial multilayer graphene. J. Phys. Condens. Matter **20**, 323202 (2008)
3. Li, X., Cai, W., An, J., Kim, S., Nah, J., Yang, D., Piner, R., Velamakanni, A., Jung, I., Tutuc, E., Banerjee, S.K., Colombo, L., Ruoff, R.S.: Large-area synthesis of high-quality and uniform graphene films on copper foils. Science **324**, 1312–1314 (2009)
4. Ding, X., Ding, G., Xie, X., Huang, F., Jiang, M.: Direct growth of few layer graphene on hexagonal boron nitride by chemical vapor deposition. Carbon **49**, 2522–2525 (2011)
5. Su, C.-Y., Lu, A.-Y., Wu, C.-Y., Li, Y.-T., Liu, K.-K., Zhang, W., Lin, S.-Y., Juang, Z.-Y., Zhong, Y.-L., Chen, F.-R., Li, L.-J.: Direct formation of wafer scale graphene thin layers on insulating substrates by chemical vapor deposition. Nano Lett. **11**, 3612–3616 (2011)
6. Kotakoski, J., Meyer, J.C.: Mechanical properties of polycrystalline graphene based on a realistic atomistic model. Phys. Rev. B **85**, 195447 (2012)
7. Kim, K.S., Zhao, Y., Jang, H., Lee, S.Y., Kim, J.M., Kim, K.S., Ahn, J.-H., Kim, P., Choi, J.-Y., Hong, B.H.: Large-scale pattern growth of graphene films for stretchable transparent electrodes. Nature **457**, 706–710 (2009)
8. Cai, W., Moore, A.L., Zhu, Y., Li, X., Chen, S., Shi, L., Ruoff, R.S.: Thermal transport in suspended and supported monolayer graphene grown by chemical vapor deposition. Nano Lett. **10**, 1645–1651 (2010)
9. Novoselov, K.S., Fal'ko, V.I., Colombo, L., Gellert, P.R., Schwab, M.G., Kim, K.: A roadmap for graphene. Nature **490**:192–200 (2012)
10. Balandin, A.A., Ghosh, S., Bao, W., Calizo, I., Teweldebrhan, D., Miao, F., Lau, C.N.: Superior thermal conductivity of single-layer graphene. Nano Lett. **8**, 902–907 (2008)
11. Seol, J.H., Jo, I., Moore, A.L., Lindsay, L., Aitken, Z.H., Pettes, M.T., Li, X., Yao, Z., Huang, R., Broido, D., Mingo, N., Ruoff, R.S., Shi, L.: Two-dimensional phonon transport in supported graphene. Science **328**, 213–216 (2010)
12. Nika, D.L., Balandin, A.A.: Two-dimensional phonon transport in graphene. J. Phys. Condens. Matter **24**, 233203 (2012)

13. Balandin, A.A.: Thermal properties of graphene and nanostructured carbon materials. Nat. Mater. **10**, 569–581 (2011)
14. Roscoe, C., Thomas, J.M.: The revelation of small-angle boundaries and forest dislocations in graphite monocrystals. Carbon **4**, 383–390 (1966)
15. Albrecht, T.R., Mizes, H.A., Nogami, J., S-i, Park, Quate, C.F.: Observation of tilt boundaries in graphite by scanning tunneling microscopy and associated multiple tip effects. Appl. Phys. Lett. **52**, 362 (1988)
16. Červenka, J., Flipse, C.F.J.: Structural and electronic properties of grain boundaries in graphite: Planes of periodically distributed point defects. Phys. Rev. B **79**, 195429 (2009)
17. Kim, K., Lee, Z., Regan, W., Kisielowski, C., Crommie, M.F., Zettl, A.: Grain boundary mapping in polycrystalline graphene. ACS Nano **5**, 2142–2146 (2011)
18. Koepke, J.C., Wood, J.D., Estrada, D., Ong, Z.-Y., He, K.T., Pop, E., Lyding, J.W.: Atomic-scale evidence for potential barriers and strong carrier scattering at graphene grain boundaries: a scanning tunneling microscopy study. ACS Nano **7**, 75–86 (2013)
19. Liu, T.-H., Gajewski, G., Pao, C.-W., Chang, C.-C.: Structure, energy, and structural transformations of graphene grain boundaries from atomistic simulations. Carbon **49**, 2306–2317 (2011)
20. Zhang, J., Zhao, J.: Structures and electronic properties of symmetric and nonsymmetric graphene grain boundaries. Carbon **55**, 151–159 (2013)
21. Cockayne, E., Rutter, G.M., Guisinger, N.P., Crain, J.N., First, P.N., Stroscio, J.A.: Grain boundary loops in graphene. Phys. Rev. B **83**, 195425 (2011)
22. Capasso, A., Placidi, E., Zhan, H.F., Perfetto, E., Bell, J.M., Gu, Y.T., Motta, N.: Graphene ripples generated by grain boundaries in highly ordered pyrolytic graphite. Carbon **68**, 330–336 (2014)
23. Hahn, K.R., Melis, C., Colombo, L.: Thermal transport in nanocrystalline graphene investigated by approach-to-equilibrium molecular dynamics simulations. Carbon **96**, 429–438 (2016)
24. Mortazavi, B., Pötschke, M., Cuniberti, G.: Multiscale modeling of thermal conductivity of polycrystalline graphene sheets. Nanoscale **6**, 3344–3352 (2014)
25. Liu, H.K., Lin, Y., Luo, S.N.: Grain boundary energy and grain size dependences of thermal conductivity of polycrystalline graphene. J. Phys. Chem. C **118**, 24797–24802 (2014)
26. Hahn, K.R., Melis, C., Colombo, L.: Structural, vibrational, and thermal properties of nanocrystalline graphene in atomistic simulations. J. Phys. Chem. C **120**, 3026–3035 (2016)
27. Melis, C., Dettori, R., Vandermeulen, S., Colombo, L.: Calculating thermal conductivity in a transient conduction regime: theory and implementation. Eur. Phys. J. B **87**, 96 (2014)
28. Yazyev, O.V., Louie, S.G.: Electronic transport in polycrystalline graphene. Nat. Mater. **9**, 806–809 (2010)
29. Plimpton, S.: Fast parallel algorithms for short-range molecular dynamics. J. Comput. Phys. **117**, 1–19 (1995)
30. Plimpton, S., Thompson, A., Crozier, P.: LAMMPS Molecular Dynamics Simulator. http://lammps.sandia.gov
31. Barbarino, G., Melis, C., Colombo, L.: Effect of hydrogenation on graphene thermal transport. Carbon **80**, 167–173 (2014)
32. Barbarino, G., Melis, C., Colombo, L.: Intrinsic thermal conductivity in monolayer graphene is ultimately upper limited: a direct estimation by atomistic simulations. Phys. Rev. B **91**, 035416 (2015)
33. Malola, S., Häkkinen, H., Koskinen, P.: Structural, chemical, and dynamical trends in graphene grain boundaries. Phys. Rev. B **81**, 165447 (2010)
34. Lahiri, J., Lin, Y., Bozkurt, P., Oleynik, I.I., Batzill, M.: An extended defect in graphene as a metallic wire. Nat. Nanotechnol. **5**, 326–329 (2010)
35. Helgee, E.E., Isacsson, A.: Scattering of flexural acoustic phonons at grain boundaries in graphene. Phys. Rev. B **90**, 045416 (2014)
36. Liu, X., Zhang, G., Zhang, Y.-W.: Thermal conduction across graphene cross-linkers. J. Phys. Chem. C **118**, 12541–12547 (2014)

37. Ferrari, A.C.: Raman spectroscopy of graphene and graphite: Disorder, electron–phonon coupling, doping and nonadiabatic effects. Solid State Commun. **143**, 47–57 (2007)
38. Bissett, M.A., Tsuji, M., Ago, H.: Strain engineering the properties of graphene and other two-dimensional crystals. Phys. Chem. Chem. Phys. **16**, 11124–11138 (2014)
39. Mohiuddin, T., Lombardo, A., Nair, R., Bonetti, A., Savini, G., Jalil, R., Bonini, N., Basko, D., Galiotis, C., Marzari, N., Novoselov, K., Geim, A., Ferrari, A.: Uniaxial strain in graphene by Raman spectroscopy: G peak splitting, Grüneisen parameters, and sample orientation. Phys. Rev. B **79**, 205433 (2009)
40. Hahn, K.R., Puligheddu, M., Colombo, L.: Thermal boundary resistance at Si/Ge interfaces determined by approach-to-equilibrium molecular dynamics simulations. Phys. Rev. B **91**, 195313 (2015)
41. Schelling, P.K., Phillpot, S.R., Keblinski, P.: Comparison of atomic-level simulation methods for computing thermal conductivity. Phys. Rev. B **65**, 144306 (2002)
42. Sellan, D.P., Landry, E.S., Turney, J.E., McGaughey, A.J.H., Amon, C.H.: Size effects in molecular dynamics thermal conductivity predictions. Phys. Rev. B **81**, 214305 (2010)
43. Yasaei, P., Fathizadeh, A., Hantehzadeh, R., Majee, A.K., El-Ghandour, A., Estrada, D., Foster, C., Aksamija, Z., Khalili-Araghi, F., Salehi-Khojin, A.: Bimodal phonon scattering in graphene grain boundaries. Nano Lett. **15**, 4532–4540 (2015)

Raman Spectroscopy of Graphene Nanoribbons: A Review

C. Casiraghi and D. Prezzi

Abstract In the last few years, several methods have been proposed for the production of ultra-narrow stripes of graphene, called graphene nanoribbons, which could find several applications in nano- and opto-electronics. However, every production method gives rise to different types of ribbons, in terms of structural quality, width, edge pattern, and type of functional groups. In this review, we compare the Raman spectrum of graphene nanoribbons produced by different techniques by focusing on the effect of different structural parameters, such as width and edge-patterns. Effects due to changes in the lateral dimension are also discussed by comparing the Raman spectrum of ultra-narrow graphene nanoribbons with the spectrum of polycyclic aromatic hydrocarbons and defective graphene.

1 Introduction

The outstanding properties of graphene make it a very attractive material for many applications, ranging from electronics to energy and health care [1]. In particular, graphene exhibits ambipolar field effect and has high-mobility charge carriers of unique relativistic nature [2]. However, the lack of a bandgap strongly restricts the use of graphene in field-effect transistors and opto-electronics. Graphene nanoribbons (GNRs), defined as nanometer-wide stripes of graphene, are expected to combine the unique properties of graphene with the presence of a finite bandgap, produced by quantum confinement [3]. Quantum mechanics predicts the gap of a material to inversely increase with its lateral dimension [3], which in the case of GNRs corresponds to their width (w). However, calculations have demonstrated that the GNRs' electronic properties strongly depend on both the width, sometimes

C. Casiraghi (✉)
School of Chemistry, University of Manchester, Manchester, UK
e-mail: cinzia.casiraghi@manchester.ac.uk

D. Prezzi (✉)
CNR-Nanoscience Institute, S3 Center, Modena, Italy
e-mail: deborah.prezzi@nano.cnr.it

V. Morandi and L. Ottaviano (eds.), *GraphITA*, Carbon Nanostructures,
DOI 10.1007/978-3-319-58134-7_2

defined as multiple of the N carbon dimer lines across the width-axis direction, and the specific atomic arrangement at the edges [4, 5]. The most studied edges are those obtained by following high-symmetry directions in the graphene lattice, i.e., zigzag (Z) and armchair (A) edges.

GNRs with well-defined physical properties, and therefore suitable for nano-electronics, are required to be very narrow ($w \ll 10$ nm) and atomically precise, in particular at the edges [4, 5]. However, the production of such GNRs is very challenging. Standard top-down approaches (e.g. litography, etching, cutting of graphene) yield GNRs with moderate width (>10 nm) and provides only relatively poor control over the edges [6–11]. Since the atomic structure of GNR reproduces an unrolled single-walled carbon nanotube (CNT), chemical unzipping of CNTs has been widely used for GNR's production [12–22], but only very few works reported ribbons with width below 10 nm [12, 14]. Alternative methods rely on local probe lithography [23], and solution methods [24–26]. Epitaxial-based methods have been also reported [27–33], such as plasma and chemical vapor deposition (PCVD and CDV, respectively). Recently bottom-up techniques based on surface-assisted [34–36] or on solution-mediated [37–43] cyclo-dehydrogenation have enabled the production of ultra-narrow ($w \ll 10$ nm) and atomically precise GNRs. The main advantage of the bottom-up methods compared to other techniques relies on the ability to determine the precise ribbons geometry by designing the structure of the molecular precursors on purpose. Note that in the case of solution-mediated GNRs the edges are typically functionalized by alkyl chains to provide a better solubility and to keep the ribbons planar [40].

Every production method gives rise to ribbons with very different structures, in term of width, length, edge morphology, position, and type of functional groups. Therefore, characterization is crucial. Raman spectroscopy is a well-established, fast, and non-destructive characterization technique for various carbon nanomaterials, including diamond-like carbons, CNTs, polycyclic aromatic hydrocarbons (PAHs) and graphene [44–46]. It is therefore expected that Raman spectroscopy could be used to fully characterize GNRs, as done with graphene. However, dimension is one of the most defining materials' parameter and we still lack a full understanding of how electrons, vibrations, and their interactions are affected by the change in the lateral dimension. Given the usually large length to width ratio, these systems can be regarded as infinite quasi-one-dimensional materials, and many similarities with graphene and CNTs are expected. On the other hand, the finite lateral size and the presence of edges introduce a molecular character, which allows making analogies with their finite counterparts, i.e., PAHs. Also, for sufficiently small length, the effect of three-dimensional confinement would make GNRs behave as elongated aromatic molecules. As such, GNRs offer a unique opportunity to study the effect of different structural parameters, such as finite width and edge-patterns, and to get insights on the evolution of the vibrational properties with the lateral dimension.

Here we report a mini-review on Raman spectroscopy of GNRs. Despite the importance of lattice vibrations for characterization purposes and transport applications, there have been relatively few works on this topic as compared to the case of electronic properties. From an experimental point of view, most of the studies are devoted to the characterization of the high-energy region to recognize the presence of sp^2 carbon networks. Nonetheless, many works on GNRs do not report any Raman spectrum, making it difficult to compare the quality of GNRs produced by different methods. Another complication in comparing the literature results is related to the use of different experimental conditions, such as the laser wavelength, which is well known to strongly influence the Raman spectrum of sp^2-based carbon nanostructures [44, 45, 47, 48]. From the theory side, most of the works have been instead devoted to the investigation of vibrations in ideal armchair and zigzag GNRs of nanometric width [49–57], mainly focusing on edge-related modes in different energy range. For long a direct comparison with theory was mostly inhibited due to the edge roughness and the larger widths of the samples. Only recently, after the production of atomically precise GNRs from bottom-up techniques, few combined theoretical and experimental works appeared, trying to explain the observed experimental features [34, 58, 59].

2 High-Energy Spectral Region

Figure 1 shows a selection of Raman spectra measured for ribbons produced by different methods, while Table 1 compares different Raman data reported in the literature. Note that for a meaningful comparison, we selected only measurements performed in the same excitation energy range, i.e., 514–532 nm, which are the most used laser wavelengths.

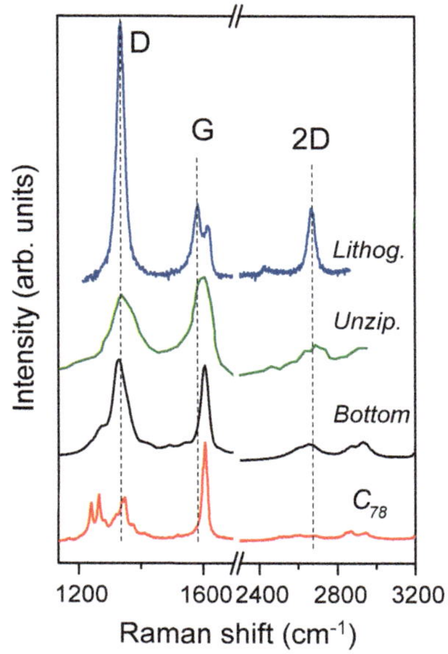

Fig. 1 Representative Raman spectra of GNRs produced with different methods: lithography (~30 nm width) [7]; unzipping (~2 nm width) [12] and bottom up (~1 nm width) [40]. The Raman spectrum of C_{78} is also reported for comparison [67]. All spectra have been measured with an excitation wavelength in the range 514–532 nm

Table 1 Selection of works discussed in this review. The Raman spectrum of all GNRs is measured using an excitation wavelength in the range 514–532 nm. The third column reports the width of the ribbon (w) and every detail reported (e.g. length, thickness, quality of the edges, functional groups). The last column summarizes the main Raman spectrum features, in particular I(D)/I(G) and FWHM(D), which can be used to qualitatively distinguish the GNRs from defective graphene (see Fig. 2)

References	Method	Properties	Raman spectrum
[12]	Unzipping	$w < 5$ nm Edges: "Relatively smooth"	I(D)/I(G) ~ 0.75 FWHM(D) ~ 100 cm^{-1}
[13]	Unzipping	$w = 130$–250 nm	I(D)/I(G) ~ 1 FWHM(G) ~ 110 cm^{-1}
[14]	Unzipping	$w = 2$–5 nm Edges: "mainly zigzag"	I(D)/I(G) ~ 0.53 (dependent on polarization) FWHM(D) ~ 65 cm^{-1}
[15]	Unzipping	$w \sim 100$ nm	I(D)/I(G) = 0.45 FWHM(D) = 95 cm^{-1}
[16]	Unzipping	$w = 80$–120 nm	I(D)/I(G) = 0.5–1.1 FWHM(D) = 70–120 cm^{-1}
[17]	Unzipping	$w = 20$–30 nm Oxygen-based groups	I(D)/I(G) = 0.5 FWHM(D) ~ 125 cm^{-1}
[18]	Unzipping	$w = 50$–100 (from TEM)	I(D)/I(G) = 0.7 FWHM(D) ~ 60 cm^{-1}
[7]	Lithography	$w = 30$–500 nm	I(D)/I(G) α 1/w FWHM(D) depends on w
[30]	CVD	$w = 10$–50 nm (by TEM) Length > 20 μm Sp^3 ~ 15%	I(D)/I(G) ~ 0.76 FWHM(D) = 100 cm^{-1} No 2D peak
[27]	CVD	$w = 20$–300 nm Sp^3 = 39% Thickness = 2–40 layers Edges: "relatively sharp"	I(D)/I(G) = 1.1 FWHM(D) ~ 70 cm^{-1} 2D peak visible
[33]	CVD	$w = 15$–20 nm Thickness = 3–5 Layers	I(D)/I(G) ~ 1 FWHM(D) ~ 95 cm^{-1}
[28]	CVD	$w \sim 100$ nm (by SEM)	I(D)/I(G) ~ 0.8 FWHM(D) ~ 120 cm^{-1}
[31]	Epitaxial growth on BN	$w = 15$–150 nm	I(D)/I(G) ~ 0.6 FWHM(D) ~ 20 cm^{-1}
[36]	Bottom up	$w = 1.5$–4.2 nm (by STM) Type: 7-,14-, 24-GNR Edges: Armchair	I(D)/I(G) ~ 1 FWHM(D) ~ 35 cm^{-1} RBLM at 396 cm^{-1} Peak at 2000 cm^{-1} observed
[43]	Bottom up	$w \sim 1$ nm, Type: Chevron-like Edges: Armchair	I(D)/I(G) ~ 0.55 FWHM(D) ~ 65 cm^{-1}
[34]	Bottom up	$w \ll 5$ nm Type: Chevron-like	I(D)/I(G) ~ 0.6 FWHM(D) ~ 55 cm^{-1}
[40]	Bottom up	$w \sim 1$–2 nm Type: Cove-shaped	I(D)/I(G) ~ 0.6–1 FWHM(D) ~ 50–60 cm^{-1}

In general the Raman spectrum of GNRs appears very similar to defective graphene, with typical defect-activated peaks [44–48], no matter the production method used. However, there could be some notable differences, depending on the GNR quality and precise structure. The first-order Raman spectrum of graphene consists of the D, G, 2D peaks, which are due to in-plane vibrations [44, 45, 60]. The G peak corresponds to the high frequency E_{2g} phonon at **Γ**. The D peak is due to the breathing modes of six-atom rings and requires a defect for its activation [47, 60–62]. It is activated by a resonant process [62, 63], and is strongly dispersive with excitation energy (~50 cm^{-1}/eV in graphene) [48, 62, 63]. The D' peak is activated by an intra-valley and resonant process, i.e., connecting two points belonging to the same cone around **K** (or **K'**). The 2D peak is the D peak overtone, while the 2D' peak is the D' overtone.

Let us focus on the D peak. The effect of an increasing amount of defects on the Raman spectrum of graphene can be described with a phenomenological three-stage amorphization trajectory [47, 64]. In stage 1, starting from pristine graphene, the Raman spectrum evolves as follows: the D peak appears and the intensity ratio between the D and G peaks [I(D)/I(G)] increases; the D' appears; all the peaks broaden and G and D' begin to overlap. In this stage, I(D)/I(G) can be used to estimate the amount of defects [64, 65], while I(D)/I(D') can be used to distinguish between different type of defects [66]. At the end of stage 1, the G and D' peaks are no more distinguishable, and I(D)/I(G) starts decreasing. Stage 2 shows a marked decrease in the G peak position and strong broadening of the peaks; I(D)/I(G) sharply decreases towards zero and second-order peaks are no more well defined. Stage 3 describes amorphous materials with increasing sp^3 content [47, 64, 67]. In this stage, the Raman spectrum shows an increase in the G peak position, I(D)/I(G) ratio is close to zero and the G peak becomes dispersive with the excitation energy [47, 64, 67].

Many works on GNRs comment on the quality of their samples by looking at I(D)/I(G) only. It is commonly believed that a decrease in I(D)/I(G) should correspond to better quality GNRs (i.e., less disordered edges). However, this statement is true only for defective graphene in stage 1 and under specific conditions. First, comparison between intensity ratio is valid only if the spectra are measured under the same experimental conditions (not only excitation wavelength, but also laser polarization and type of ribbons, i.e., aligned or randomly distributed ribbons, and type of edges and thickness, i.e., single-layers or multi-layers). We remark for example that ribbons with perfect zigzag edges are expected not to show any D peak, while armchair ones are expected to always display a D peak [68, 69]. Furthermore, even by assuming that the three-stage model is valid for all GNRs, then a decrease in I(D)/I(G) is not necessarily related to an increase in edge quality. Many ribbons show small I(D)/I(G), but belong to stage 2, as shown by looking at both I(D)/I(G) and the FWHM of the Raman peaks, e.g., of the D peak (FWHM(D)). Figure 2 collects I(D)/I(G) as a function of FWHM(D) for the ribbons listed in Table 1; the blue shadow shows the amorphization trajectory, determined for defective graphene and graphene oxide (GO) [70]. Figure 2 shows that GO GNRs [26] belong to stage 2, very close to GO, as expected from to their strongly disordered nature. Furthermore, it is interesting to observe that for lithography produced ribbons with no controlled edges the cross-over between stage 1 and 2 is observed at ~30 nm width. Therefore, a ribbon with larger width, but belonging to stage 2 is expected to contain disorder that is not only localized at the edges. This is the case of the GNRs produced in ref 27: from the Raman spectrum I(D)/I(G) = 1.1 and FWHM(D) ~70 cm^{-1}, therefore the GNRs may belong to stage 2, although they are less disordered than GO GNRs. This is in agreement with a measured sp^3 content of 39% [27]. Figure 2 also shows that methods based on CNTs unzipping can produce GNRs of very different quality, depending on the type of nanotubes and specific unzipping method used. Based on their Raman spectra, some of those GNRs belong to stage 2, and contain a relatively high amount of disorder, at least compared to GNRs of similar width produced by lithography, which are still in stage 1. Other GNRs, produced by unzipping, show very different Raman spectra, which do not belong to the amorphization trajectory. The same is observed for all GNRs produced by bottom-up approaches. This is a signature of confinement effect, which strongly changes the Raman spectrum and do not allow using the amorphization trajectory for the analysis of those nanostructures.

Let us take for comparison the Raman spectrum of an elongated aromatic molecule, such as $C_{78}H_{26}$ (C_{78}) [71]. The Raman spectrum of C_{78} shows Raman modes around 1600 and 1300 cm^{-1}, which are associated to the same eigenvectors and displacement patterns of the peaks corresponding to the G and D bands in graphene [72]. Note that unlike the case of graphene, where the D peak requires a defect for its activation, in aromatic molecules the vibrations corresponding to the D

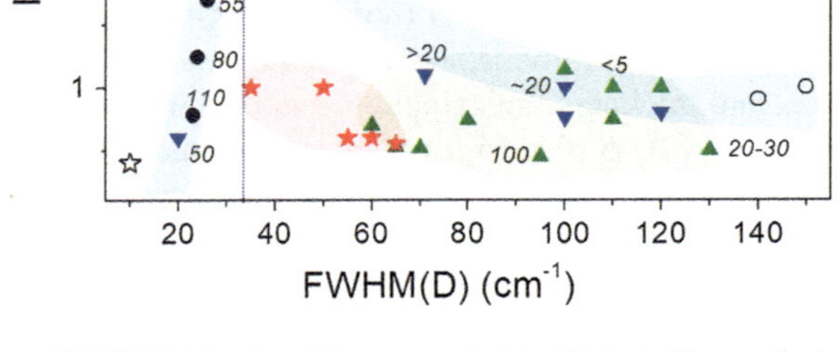

Fig. 2 I(D)/I(G) vs FWHM(D) for the ribbons reported in Table 1. The vertical dotted line marks the over-crossing between stage 1 and stage 2. The *blue* shadow represents the amorphization trajectory, starting from graphene and ending with GO. The *red* and *green* shadows collect the data for bottom-up and unzipped GNRs, respectively. The numbers close to the points refer to the width of the GNRs in nm. For in C_{78} and bottom-up ribbons we considered the most intense contribution of the D peak

peak are Raman active and do not require any defect to be seen [72]. In contrast to defective graphene, the Raman spectrum of C_{78} is characterized by several sharp (FWHM ~10 cm^{-1}) vibrational peaks in the region 1300–1400 cm^{-1}, Fig. 1, associated to different ring vibrations [71]. The intensity of these modes is also strongly dominated by resonance effects due to the presence of a gap between the LUMO and HOMO levels [71]. This is very different from the D peak resonance-activated mode in graphene, where the resonance effects are caused by the unique electronic and phonon structure of graphene. As mentioned above, all GNRs produced by bottom-up methods do not belong to the graphene amorphization trajectory and typically show a structured D peak, with a prominent component, which reminds of the C_{78} Raman spectrum, Fig. 1. This is a signature of the peculiar character of GNRs, which are quasi-infinite in the length direction, but finite in the width direction, giving rise to fundamental and overtone modes, as theoretically predicted in Refs. [55–57]. This was demonstrated for both AGNRs [34, 59] and cove-shaped GNRs [58], from the comparison with ab initio density functional theory (DFT) calculations. The Raman characterization of ZGNRs, recently obtained from surface-assisted bottom-up approaches [35], is instead still absent.

Remarkably, Yamada et al. [55] investigated the phonon dispersion of A- and ZGNRs from zone folding (ZF) by using the force constants of PAH based on force-field methods, while Gillen et al. [56, 57] performed ab initio DFT calculations and compare their results unfolding the GNR modes onto the Brillouin zone of graphene. Both these approaches are useful to evidence differences and similarities with PAHs and graphene. In contrast to graphene, the two in-plane optical modes (transverse, TO, and longitudinal, LO) are not degenerate, even though their frequency tends to converge to that of graphene for large widths. The LO-TO splitting shows a clear dependence on the ribbon width and edge type, similar to what happens for electronic [73] and optical properties [74]. The analysis of the Raman activity also shows a dependence on the same parameters [57]. The GNR overtones reproduce quite well the phonon dispersion of graphene, especially near the Γ point. Deviations are larger near the zone edge due to the finite widths of the ribbons, and near the K point due the absence of Kohn anomalies [75] induced by the gap opening in GNRs. This analysis was extremely helpful for the interpretation of experimental observations [34, 58, 59]. In addition to G- and D-like peaks, also edge-related phonon modes also raised great interest. Among the others, a transverse edge-localized Raman active mode at ~2000 cm^{-1}, typical of non-hydrogenated AGNRs, was predicted theoretically [49, 50] and detected experimentally few years later for GNRs grown on TiC(755) [76] and more recently in bottom-up GNRs [36]. Other lower energy modes localized at the edges have been reported below 1200 cm^{-1}, and their Raman and infrared activity characterized, also in comparison with PAHs [52, 53, 55].

3 Low-Energy Spectral Region

The Raman spectrum of GNRs shows important features in the low-energy region. Specific fingerprints have been observed in this spectral region for a variety of other carbon allotropes: most notably, in the case of CNTs, a characteristic peak at low energy, associated to the radial breathing mode (RBM) of all the atoms of the structure, has been widely exploited to determine the tube diameter [45, 77]. PAHs are also characterized by breathing-like low-energy modes that can be related to their lateral size [71]. Raman signatures in this spectral region have been predicted theoretically [52, 57] for GNRs, and recently observed experimentally. Reference [34] first reported a sharp peak at 396 cm^{-1} for surface-assisted GNRs, and assigned to a Radial-Like Breathing Mode (RLBM), where all the atoms of the ribbon move in plane in directions perpendicular to the ribbon width [52]. The RLBM position is expected to depend on the GNR width, in analogy to the RBM in carbon nanotubes [45, 77]. Gillen et al. [57] calculated from first principles the frequency of the RLBM for a number of A- and ZGNRs of increasing width, mapping it onto the LA branch of graphene. The agreement, especially for sufficiently large ribbons ($w > 1.0$–1.5 nm), allows to extract an expression for the RLBM frequency based on zone folding: $\nu_{RBLM} = 3222/w$, where ν_{RBLM} is the RBLM position in cm^{-1} and the width is in Å [57]. Considering that the cut of the edge filter in standard spectrometers is around 150 cm^{-1}, ribbons with width below ~2.2 nm should show a RLBM mode. Currently, this mode has been observed only in the Raman spectrum of bottom-up fabricated ribbons [34, 36, 40, 41, 58].

Because bottom-up approaches can generate ribbons with different geometry (e.g., cove-shaped, chevron-like, etc.), one has to be careful in extending the relation between width and RLBM position to all GNRs. One of our recent works [58] demonstrated that the relation between width and RLBM position can become rather complicated for cove-shaped GNRs, where a benzo ring periodically decorates the zigzag edge. In these GNRs the RLBM is not located exactly at the position expected using the equation above [58]. Furthermore, the RLBM does not appear as a single and sharp mode, as observed for the RBM in CNTs, but it is a broad peak, often composed by several overlapping components [58]. A number of other peaks with small intensity are also seen in the spectral range 400–700 cm^{-1} [58]. The RLBM peak is also slightly dispersive, although the dispersion changes with the width and edge pattern of the ribbon, in similar fashion to the D peak [58]. In general the Raman spectrum of cove-shaped GNRs is strongly sensitive to the exact edge pattern of the GNR [58]. Therefore, ab initio simulations have been very useful to understand how every single detail of the ribbons structure can influence the Raman features. These show that a better estimate for the RLBM frequency can be found by applying the above expression with an effective width to take into account the width modulation in cove-shaped GNRs, Fig. 3. For example for the cove-shaped 8GNR with hydrogenated edges (+H), the RLBM position is in between a 8Z-GNR and 10Z-GNR. By using the effective width, calculated as the weighted average of the different GNR widths, a good agreement between first

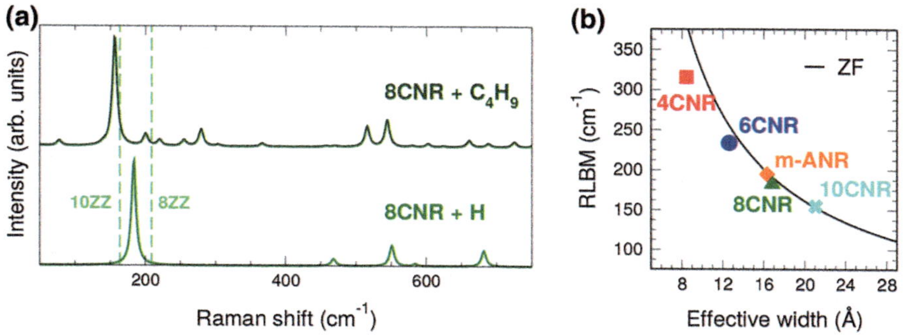

Fig. 3 **a** Simulated acoustic region of the Raman spectrum of cove-shaped carbon nanoribbons (CNRs) with N = 8. The spectrum is shown for both hydrogen-terminated (+H, *green*) and functionalized 8CNR (+C_4H_9, *dark green*). The *dashed lines* indicate the position of the RLBM for 8- and 10-ZGNRs (labeled 8-ZZ and 10-ZZ, respectively, *light green*) and the position of the G peak for 8-ZGNR. **b** Frequency of the RLBM calculated from first principles for several H-passivated cove-shaped GNRs is compared to the result of the zone folding (ZF) approximation (*black curve*) as a function of the GNR effective width. Adapted from Ref. [55]

principles simulations and the results obtained by zone folding is recovered [58], Fig. 3b. The higher frequency modes at 470–680 cm^{-1}, Fig. 3a, combining longitudinal and transverse components and not present in ZGNRs, appear in cove-shaped GNRs in view of the different periodicity along the ribbon axis introduced by the additional benzo rings at the edge. The effect of the side chains functionalizing the GNRs to improve their solubility has been also investigated by first principles calculations. Depending on the length of the alkyl chain and on the GNR width, the RLBM can be significantly modified, coupling to the modes of the chain and giving rise to several subpeaks, Fig. 3a. This can also explain the broadening of the peak observed experimentally [58].

4 Conclusions

The Raman spectrum of graphene nanoribbons is very complex due to the dual nature of GNRs, which are quasi-infinite in the length direction, but finite in the width direction. Therefore, GNRs with different width, edge pattern, and amount of defects can show very different Raman spectra. This makes Raman spectroscopy a very powerful tool for comparing different production techniques, although a systematic study, where a large variety of GNRs, produced by different methods, are studied under the same experimental conditions is currently still missing. Raman measurements on individual GNRs would be also useful to avoid any effect due to stacking; this type of sample would also allow performing resonant Raman spectroscopy, which could provide further information on electronic states and excitons in such nanostructures.

Acknowledgements The authors acknowledge A.C. Ferrari for useful discussions; I. Verzibisky, M. Tommasini, C. Stampfer, and D. Bischoff for providing some of the Raman spectra in Fig. 1.

References

1. Novoselov, K.S., Falko, V.I., Colombo, L., Gellert, P.R., Schwab, M.G., Kim, K.: Nature **490**, 192–200 (2012)
2. Geim, A.K., Novoselov, K.S.: Nature Mater. **6**, 183–191 (2007)
3. Torres, L.E.F., Roche, S., Charlier, J.-C.: Introduction to Graphene-Based Nanomaterials: From Electronic Structure to Quantum Transport. Cambridge University press (2014)
4. Nakada, K., Fujita, M., Dresselhaus, G., Dresselhaus, M.S.: Phys. Rev. B **54**, 17954 (1996)
5. Wakabayashi, K., Fujita, M., Ajiki, H., Sigrit, M.: Phys. Rev. B **59**, 8271 (1999)
6. Han, M., Ozylmaz, B., Zhang, Y., Kim, P.: Phys. Rev. Lett. **98**, 206805 (2007)
7. Bischoff, D., Güttinger, J., Dröscher, S., Ihn, T., Ensslin, K., Stampfer, C.: J. Appl. Phys. **109**, 073710 (2011)
8. Ryu, S., Maultzsch, J., Han, M.Y., Kim, P., Brus, L.E.: ACS Nano **5**, 4123–4130 (2011)
9. Wang, X., Dai, H.: Nature Chem. **2**, 661 (2010)
10. Campos, L.C., Manfrinato, V.R., Sanchez-Yamagishi, J.D., Kong, J., Jarillo-Herrero, P.: Nano Lett. **9**, 2600 (2009)
11. Datta, S.S., Strachan, D.R., Khamis, S.M., Johnson, A.T.C.: Nano Lett. **8**, 1912 (2008)
12. Wei, D., Xie, L., Lee, K.K., Hu, Z., Tan, S., Chen, W., Sow, C.H., Chen, K., Liu, Y., Wee, A. T.S.: Nat. Commun. **4**, 1374 (2013)
13. Kosynkin, D.V., Lu, W., Sinitskii, A., Pera, G., Sun, Z., Tour, J.M.: ACS Nano **5**, 968 (2011)
14. Gong, Y., et al.: Phys Rev B **87**, 165404 (2013)
15. Shimizu, T., Haruyama, J., Marcano, D.C., Kosinkin, D.V., Tour, J.M., Hirose, K., Suenaga, K.: Nat. Nano **6**, 45–50 (2011)
16. Li, Y.-S., Liao, J.-L., Wang, S.-Y., Chiang, W.-H.: Sci. Rep. **6**, Article number: 22755 (2016)
17. Cataldo, F., et al.: Carbon **48**, 2596 (2010)
18. Cano-Marquez, A.G., et al.: Nano Lett. **9**, 1527 (2009)
19. Jiao, L., Wang, X., Diankov, G., Wang, H., Dai, H.: Nat. Nanotechnol. **5**, 321 (2010)
20. Jiao, L., Zhang, L., Wang, X., Diankov, G., Dai, H.: Nature **458**, 877 (2009)
21. Elías, A.L., Botello-Méndez, A.R., Meneses-Rodríguez, D., Jehová González, V., Ramírez-González, D., Ci, L., Muñoz-Sandoval, E., Ajayan, P.M., Terrones, H., Terrones, M., Nano Lett. **10**, 366 (2010)
22. Zhu, Y., Tour, J.M.: Nano Lett. **10**, 4356 (2010)
23. Magda, G.Z., Jin, X., Hagymási, I., Vancsó, P., Osváth, Z., Nemes-Incze, P., Hwang, C., Biró, L.P., Tapasztó, L.: Nature **514**, 608 (2015)
24. Li, X., Wang, X., Zhang, L., Lee, S., Dai, H.: Science **319**, 1229 (2008)
25. Ren, X.D., Liu, R., Zheng, L.M., Ren, Y.P., Hu, Z.Z., He, H.: Appl. Phys. Lett. **108**, 071904 (2016)
26. Higginbotham, A.L., Kosynkin, D.V., Sinitskii, A., Sun, Z., Tour, J.M.: ACS Nano **4**, 2059 (2010)
27. Campos-Delgado, J., et al.: Nano Lett. **8**, 2773 (2008)
28. Zhang, J., et al.: J. Alloys and Comp. **649**, 933 (2015)
29. Suzuki, H., et al.: Nature. Commun. **7**, 11797 (2016)
30. Sokolov, A.N., et al.: Nature. Commun. **4**, 2402 (2013)
31. Lu, X., et al.: Appl. Phys. Lett. **108**, 113103 (2016)
32. Jacobberger, R.M., et al.: Nature Comms **6**, 68006 (2015)
33. Martin-Fernandez, I., Wang, D., Zhang, Y.: Nano Lett. **12**, 6175 (2012)
34. Cai, J., Ruffieux, P., Jaafar, R., Bieri, M., Braun, T., Blankenburg, S., Muoth, M., Seitsonen, A.P., Saleh, M., Feng, X., et al.: Nature **466**, 470–473 (2010)

35. Ruffieux, P., Wang, S., Yang, B., Sánchez-Sánchez, C., Liu, J., Dienel, T., Talirz, L., Shinde, P., Pignedoli, C.A., Passerone, D., Dumslaff, T., Feng, X., Müllen, K., Fasel, R.: Nature **531**, 489–492 (2016)
36. Huang, H., et al.: Scientific Reports **2**, 983 (2012)
37. Fogel, Y., Zhi, L., Rouhanipour, A., Andrienko, D., Rader, H.J., Mullen, K.: Macromolecules **42**, 6878 (2009)
38. Dossel, L., Gherghel, L., Feng, X., Mullen, K.: Angew. Chem. Int. Ed. **50**, 2540 (2011)
39. Schwab, M.G., Narita, A., Hernandez, Y., Balandina, T., Mali, K.S., De Feyter, S., Feng, X., Muellen, K.: J. Am. Chem. Soc. **134**, 18169 (2012)
40. Narita, A., Feng, X., Hernandez, Y., Jensen, S.A., Bonn, M., Yang, H., Verzhbitskiy, I.A., Casiraghi, C., Hansen, M.R., Koch, A.H.R., Fytas, G., Ivasenko, O., Li, B., Mali, K.S., Balandina, T., Mahesh, S., De Feyter, S., Muellen, K.: Nat. Chem. **6**, 126 (2014)
41. Narita, A., Verzhbitskiy, I.A., Frederickx, W., Mali, K.S., Jensen, S.A., Hansen, M.R., Bonn, M., De Feyter, S., Casiraghi, C., Feng, X., Müllen, K.: ACS Nano **8**, 11622 (2014)
42. Gao, J., Uribe-Romo, F.J., Saathoff, J.D., Arslan, H., Crick, C.R., Hein, S.J., Itin, B., Clancy, P., Dichtel, W.R., Loo, Y.-L.: ACS Nano **10**, 4847 (2016)
43. Vo, T., Shekhirev, M., Kunkel, D.A., Morton, M.D., Berglund, E., Kong, L., Wilson, P.M., Dowben, P.A., Enders, A., Sinitskii, A.: Nat. commun. **5**, 3189 (2014)
44. Ferrari, A.C., Robertson, J.: Raman spectroscopy of amorphous, nanostructured, diamond-like carbon, and nanodiamond. Phil. Trans. R. Soc. Lond. A **362**, 2477 (2004)
45. Jorio, A., Saito, R., Dresselhaus, G., Dresselhaus, M.S.: Raman Spectroscopy in Graphene Related Systems. Wiley-VCH Verlag GmbH (2011)
46. Casiraghi, C.: Raman spectroscopy of graphene. In: Spectroscopic Properties of Inorganic and Organometallic Compounds: Techniques, Materials and Applications, vol. 43, pp. 29–56. The Royal Society of Chemistry (2012)
47. Ferrari, A.C., Robertson, J.: Phys. Rev. B **64**, 075414 (2001)
48. Eckmann, A., Felten, A., Verzhbitskiy, I., Davey, R., Casiraghi, C.: Phys. Rev. B **88**, 35426 (2013)
49. Igami, M., Fujita, M., Mizuno, S.: Appl. Surf. Sci. **130–132**, 870 (1998)
50. Kawai, T., Miyamoto, Y., Sugino, O., Koga, Y.: Phys. Rev. B **62**, R16349 (2000)
51. Yamamoto, T., Watanabe, K., Mii, K.: Phys. Rev. B **70**, 245402 (2004)
52. Zhou, J., Dong, J.: Appl. Phys. Lett. **91**, 173108 (2007)
53. Vandescuren, M., Hermet, P., Meunier, V., Henrard, L., Lambin, Ph: Phys. Rev. B **78**, 195401 (2008)
54. Malola, S., Hakkinen, H., Koskinen, P.: Eur. Phys. J. D **52**, 71 (2009)
55. Yamada, M., Yamakita, Y., Ohno, K.: Phys. Rev. B **77**, 054302 (2008)
56. Gillen, R., Mohr, M., Maultzsch, J., Thomsen, C.: Phys. Rev. B **80**, 155418 (2009)
57. Gillen, R., Mohr, M., Maultzsch, J.: Phys. Rev. B **81**, 205426 (2010)
58. Verzhbitskiy, I., et al.: Nano Lett. **16**, 3442 (2016)
59. Yang, W., Lucotti, A., Tommasini, M., Chalifoux, W.A.: J. Am. Chem. Soc. **138**, 9137–9144 (2016)
60. Tuinstra, F., Koenig, J.L.: J. Chem. Phys. **53**, 1126–1130 (1970)
61. Ferrari, A.C., Robertson, J.: Phys. Rev. B **61**, 14095–14107 (2000)
62. Thomsen, C., Reich, S.: Phys. Rev. Lett. **85**, 5214–5217 (2000)
63. Baranov, A.V., et al.: Opt. Spectroscopy **62**, 612–616 (1987)
64. Cançado, L.G., et al.: Nano Lett. **11**, 3190–3196 (2011)
65. Martins Ferreira, E.H., Moutinho, M.V.O., Stavale, F., Lucchese, M.M., Capaz, R.B., Achete, C.A., Jorio, A.: Phys. Rev. B **82**, 125429 (2010)
66. Eckmann, A., Felten, A., Mishchenko, A., Britnell, L., Krupke, R., Novoselov, K.S., Casiraghi, C.: Nano Lett. **12**, 3925 (2012)
67. Casiraghi, C., Ferrari, A.C., Robertson, J.: Phys. Rev. B **72**, 085401 (2005)
68. Casiraghi, C., Hartschuh, A., Qian, H., Piscanec, S., Georgi, C., Fasoli, A., Novoselov, K.S., Basko, D.M., Ferrari, A.C.: Nano Lett. **9**, 1433 (2009)
69. Cançado, L.G., et al.: Phys. Rev. Lett. **93**, 247401 (2004)

70. Stankovich, S., et al.: Carbon **45**, 1558 (2007)
71. Maghsoumi, A., et al.: J. Raman Spectrosc. **46**, 757 (2015)
72. Castiglioni, C., Tommasini, M., Zerbi, G.: Raman spectroscopy of polyconjugated molecules and materials: confinement effect in one and two dimensions. Phil. Trans. R. Soc. Lond. A **362**, 2425–2459 (2004)
73. Son, Y.-W., Cohen, M.L., Louie, S.G.: Phys. Rev. Lett. **97**, 216803 (2006)
74. Prezzi, D., Varsano, D., Ruini, A., Marini, A., Molinari, E.: Phys. Rev. B **77**, 041404(R) (2008)
75. Piscanec, S., Lazzeri, M., Mauri, F., Ferrari, A.C., Robertson, J.: Phys. Rev. Lett. **93**, 185503 (2004)
76. Tanaka, T., Tajima, A., Moriizumi, R., Hosoda, M., Ohno, R., Rokuta, E., Oshima, C., Otani, S.: Solid State Commun. **123**, 33 (2002)
77. Rao, A.M., Richter, E., Bandow, S., Chase, B., Eklund, P.C., Williams, K.A., Fang, S., Subbaswamy, K.R., Menon, M., Thess, A., Smalley, R.E., Dresselhaus, G., Dresselhaus, M. S.: Science **275**, 187 (1997)

Electron–Phonon Coupling in Two-Dimensional Superconductors: Doped Graphene and Phosphorene

G. Profeta, C. Tresca and A. Sanna

Abstract The advent of two-dimensional materials with the possibility to vary their physical properties by means of doping, strain, electric, and magnetic fields allows to explore novel physical effects in the two-dimensional limit, where electronic, magnetic, and structural properties can be very different with respect to three-dimensional case. For example, the possibility to synthesize a two-dimensional superconductor will open the doors to new and unexplored applications in present nanotechnology. In this respect, reliable predictions of the superconducting critical temperature from first-principles and in real materials are important prerequisite to make important advances along this line of research. In this work, we review the results of recent theoretical predictions of superconductivity in experimentally realized two-dimensional superconductors: doped graphene and doped phosphorene. And for the latter system, we also present an analysis of several realistic dopants that could induce a superconducting state.

1 Introduction

The discovery of graphene [1], the two-dimensional (2D) allotrope of carbon, has opened the era of 2D system and soon boosted the interest of nanotechnology industry for layered and 2D materials. After the successful isolation of graphene and the measurement of its extraordinary physical properties [2] many other 2D allotrope of layered 3D materials were discovered, *de-facto* establishing a completely new research line in condensed matter physics. Two-dimensional materials can be built with various techniques (molecular beam epitaxy, etching, exfoliation, chemical

G. Profeta (✉) · C. Tresca
Department of Physical and Chemical Sciences, University of L'Aquila,
Via Vetoio 10, 67100 L'aquila, Italy
e-mail: gianni.profeta@aquila.infn.it

G. Profeta · C. Tresca
SPIN-CNR, University of L'Aquila, Via Vetoio 10, 67100 L'aquila, Italy

A. Sanna
Max-Planck-Institut Für Mikrostrukturphysik, Weinberg 2, 06120 Halle, Germany

V. Morandi and L. Ottaviano (eds.), *GraphITA*, Carbon Nanostructures,
DOI 10.1007/978-3-319-58134-7_3

vapor deposition), and designed like nanostripes, junctions, arrays, thin films, etc., giving many new technological applications. One of the most attractive physical properties is represented by the possibility to tune the electron density by means of molecular or atomic dopants, electric field effect, deposition on suitable substrates, etc. As a consequence, it is possible to tune material specific properties to optimal values or induce completely new (quantum) effects exploiting the enhancing of electron correlations in 2D. For example, the possibility to have or induce superconductivity in a 2D material will add a fundamental building block in the future nanotechnology.

Indeed, in recent few years, the frontier of the research in condensed matter physics moved toward the discovery of new and unexpected superconducting (SC) 2D systems. In particular, the research lines are of three types: (i) induce a superconducting phase in a otherwise non-superconducting 2D system by means of charge doping. The most striking example is represented by lithium-doped graphene [3, 4] (see below). The role of the dopants can be twofold: vary the 2D charge density and induce extra pairing channels. (ii) Use the electronic double-layer transistor to induce a charge accumulation at the interface between (insulating) 2D material and the ionic-liquid. For example, in 2010 the group of Iwasa showed [5] able to induce superconductivity in an insulator, ZrNCl. The discovery, made possible by unprecedented technological advances, paved the way to new routes to study, manipulate, and design SC materials and devices making possible, by gating, to realize a 2D metallic state in a single surface layer of any layered semiconducting material. Indeed, the technique has been applied to many transition metal dichalcogenides (TMD), which have been proven to be superconductors with relatively high T_C: MoS_2 ($T_C \approx 10$ K) [6], WS_2 ($T_C \approx 4$ K) [7], $MoSe_2$ ($T_C \approx 7$ K) [8]. The two dimensionality of such system results in peculiar properties like, e.g., the anomalous response of the SC phase to the in-plane magnetic field, probably associated to the spin-orbit coupling and/or valley-dependent Berry curvature [9]. This research field is at the very early stage and new and unexplored phenomena, systems and effects must be highlighted. Interestingly, both the layered materials and the experimental technique are perfectly integrated with the emerging 2D technologies based on graphene and post-graphene materials. (iii) growth of inherently 2D superconducting systems. Notable and recent examples are the discovery of superconductivity in 2D Pb/Si(111) [10] and FeSe monolayer on Nb-doped $SrTiO_3$ (100) substrate [11] at a record $T_C$of 100 K [12].

The field of research is very active, considering that new and promising 2D systems are routinely discovered and synthesized. Moving from carbon (graphene) along the group IV elements the 2D form of silicon (silicene [13]), of germanium (germanene [14]) and tin (stanene [15]) with a buckled graphene-like honeycomb structure, have been observed. Moving along the nearby group V elements the most important example is represented by phosphorene [16], the single layer 2D form of black phosphorous. At the same time, very recently, the 2D form of a single group III element was reported: borophene [17].

The exploration of possible new physical properties of these 2D materials by chemical, field, and interface modification is appealing and represents the ultimate frontier of material science, for both fundamental impact and technological applications.

However, in order to guide the experimental research, propose new materials, predict novel unexpected phenomena, study possible alternative dopants and chemical modifications, a crucial ingredient is a solid, reliable and quantitative predictive, theoretical and computation framework. First-principles density functional theory (DFT) at the moment represents a formidable computational tool to predict (quantitatively) many different materials specific (normal state) properties. In the original practical formulation, expressed by the Kohn–Sham approach [18], the (Gauge breaking) superconducting phase was not explicitly included. This further step can be achieved either by means of many body perturbation theory on top of the non-superconducting Kohn–Sham system, like in the Eliashberg approach [19]; or by means of DFT for superconductors [20] (SCDFT).

Eliashberg theory is the most used theory in ab-initio superconductivity [21], in its general form it is based on the self consistent solution of a Dyson-like equations for the Nambu Green's function [22], and allows to compute the most relevant experimental observables in superconductors [23]. A large literature is devoted to its development including anisotropy, multiband effects, non-adiabaticity, and more [24–27]. However, especially to obtain a quick estimation of the superconducting T_C Eliashberg theory is often used in a parameterized form, like the one obtained by McMillan [28].

An alternative to Eliashberg theory is SCDFT [20] that is a generalization of DFT to include the effect of a superconducting symmetry breaking. First developed for phononic superconductors [29] it was more recently extended to other pairing channels [30, 31]. SCDFT has been applied to the study of elemental superconductors [32, 33] and alloys like MgB_2 [34, 35], CaC_6 [36], intercalated graphites [37], high pressure hydrides [38, 39]. At the same time, the first-principles nature of the theory and the possibility to describe real materials (not only models) allowed prediction of superconducting properties of new materials: high temperature superconductivity of hydrogen under high pressure [40–42], CaBeSi [43], potassium under pressure [44, 45].

The advantage of SCDFT with respect to Eliashberg theory is its lower computational cost and the possibility of full ab-initio inclusion of the Coulomb interaction that makes it completely parameter free. On the contrary Eliashberg's theory has the advantage of being directly formulated in terms of Feynman diagrams, therefore it does not require the cumbersome step of the functional construction that is usually the bottleneck for the development of density functional methods.

Nowadays, the superconducting critical temperature can be quantitatively predicted by first-principles DFT calculations, and can even anticipate and guide experiments in the discovery of new superconductors [46, 47]. The aim of this paper is to review first-principles DFT predictions of normal and superconducting state physical properties of new 2D materials. In order to have a widespread look of the state of the art, we review results on theoretical predictions of superconductivity in graphene, recently confirmed by the experiments on the metallization of phosphorene to induce a superconducting phase.

2 Superconducting Graphene

The possibility to turn graphene into a superconductor is a fascinating research area which involved many different approaches, both experimentally than theoretically.

A first attempt by the Morpurgo's group in 2007 [48] used a Josephson effect between graphene and superconducting electrodes which demonstrated that graphene can support supercurrents. The theoretical proposal has focused on electron correlation driven superconductivity in doped graphene [49] due to the very low electron–phonon coupling in graphene. Indeed, conventional superconductivity, although desirable, was not expected in pure graphene, mainly due to low density of states at the Fermi level on the Dirac bands and the high frequencies of the bond stretching modes. In 2012, a theoretical work, inspired by the physics of graphite intercalated compounds [3] proposed lithium doping of graphene as a possible viable way to observe a superconducting phase at relatively high temperatures. The main idea was founded on the possibility to promote the so-called interlayer band (present in pure graphene at very high energies) at the Fermi energy by means of positive potential of the ionized lithium adatoms. Indeed, at a coverage of 1 Li per 6 C atoms (LiC_6 stoichiometry), lithium induces a $\sqrt{3} \times \sqrt{3}R30°$ ($\sqrt{3}$) in-plane unit cell (with respect to a standard graphene lattice) with one adatom per unit cell, adding an additional (with respect to linear Dirac bands) band at the Fermi energy (see Fig. 1). The presence of the interlayer band enhances the electron–phonon coupling, in particular at low frequency modes where the contribution to the total electron–phonon coupling (λ) is more effective. The low energy phonons are mainly characterized by out-of-

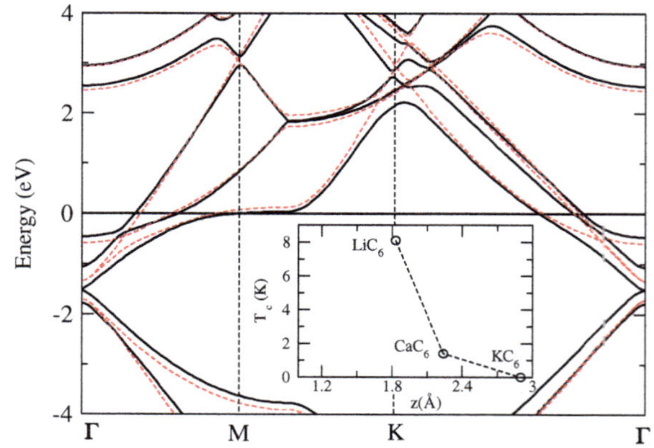

Fig. 1 Band structure of lithium-doped graphene in the $\sqrt{3}$ phase in the equilibrium configuration (*red dashed curve*) and with carbon atoms displaced along the B_{2g} phonon mode at the Γ-point (*solid black curve*)

plane displacements of carbon atoms which couples with the electrons occupying the interlayer band and localized around lithium atoms [50, 51]. Indeed, by symmetry in graphene the relevant out-of-plane phonons (B_{2g}) have the same symmetry as the π^* electronic states. Upon adatom deposition, lithium in this case, the symmetry argument is not more valid, and the coupling with B_{2g} phonons becomes possible. The electron–phonon matrix element M_i^j for the band j coupling with optical mode i can be obtained by:

$$M_j^i = \sqrt{\frac{\hbar}{2MN\omega^i}} \sum_a \epsilon_a^i \frac{\partial E_j(\mathbf{k})}{\partial \mathbf{u_a}}$$

where a denotes the atom in the unit cell, $\mathbf{k}$ and j correspond to the wavevector and band index of the electronic state. M is the atomic mass, and $\mathbf{u}_a$ is the normalized polarization eigenvector of the phonon. In order to understand the electron–phonon coupling between π states and the Γ-point B_{2g} phonons, we calculated the deformation potential, $D = \frac{\partial E_j(\mathbf{k})}{\partial \mathbf{u_a}}$, at the Γ point of π states, distorting C atoms of graphene by u along the B_{2g} phonon eigenvector and calculating of the energy shift (ΔE) between π and π^* states. We performed first-principles DFT calculation in the local density approximation [52]. The QUANTUM-ESPRESSO [53] package was used with norm-conserving pseudopotentials and a plane-wave cutoff energy of 65 Ry. All of the structures considered were relaxed to their minimum energy configuration following the internal forces on atoms and the stress tensor of the unit cell. The monolayer systems were simulated in the $\sqrt{3} \times \sqrt{3}R30°$ in-plane unit cell (with respect to a standard graphene lattice) with one adatom per unit cell [3]. In Fig. 1, we report the band structure of LiC_6 with and without the phonon deformation.

It is evident that the out-of-plane phonon mode activate a relevant splitting of both π and π^* bands giving a deformation potential of $D = 5.2\,\mathrm{eV/Å}$. Considering that the square of the electron–phonon matrix elements enters in the definition of λ it results a relevant term.

This is the most evident manifestation of the increased electron–phonon coupling in graphene, due to the adsorption of lithium adatoms. Evidently the coupling between out-of-plane B_{2g} modes depends on the distance between adatom and graphene layers. We demonstrate this aspect, calculating the superconducting critical temperature for other two prototype systems, namely: calcium on graphene (CaC_6) and potassium on graphene (KC_6). Both systems have been simulated in the same $\sqrt{3} \times \sqrt{3}R30°$ of LiC_6. The different ionic radius (r) of the adatoms ($r_{Li} = 0.8$ Å, $r_{Ca} = 1.0$ Å and $r_K = 1.4$ Å) determines a different adsorption height above the graphene layers, changing the electron–phonon coupling and the relative T_C. In the inset of Fig. 1, we report the critical temperature as a function of the equilibrium height. The calculation confirms that the critical temperature strongly increased reducing the adatom/graphene distance which influences the electron–phonon coupling with B_{2g} modes.

The theoretical proposal of a possible superconducting phase in lithium-doped graphene was recently put on a real ground by the Damascelli's group [4] which demonstrated that lithium decorated graphene presents a superconducting phase with a critical temperature of 5.9 K. The measured critical temperature is in line with the theoretical prediction giving a strong indication that the observed crystal phase is indeed the theoretical proposed $\sqrt{3} \times \sqrt{3}R30°$. One of the main indication of the formation of the superstructure is the folding of the π bands from the K-point of the graphene unit cell to the Γ-point of the $\sqrt{3}$ Brillouin zone. This last evidence ensures the presence of the interlayer state crossing the Fermi level which strongly contributes to the increase of the electron–phonon coupling [54]. In Ref. [4], the authors measured the Γ-point intensity of the spectral weight as a function of the deposition time, revealing the presence of the Li-2s band expected for this system [3] superimposed with the folded graphene bands caused by a Li superstructure, which were observed in Li and Ca bulk GIC systems [54, 55]. This spectral weight, which disappears above $\simeq$50 K and is not recovered on subsequent cooling, is associated with the strong enhancement of electron–phonon coupling.

In Fig. 3E of Ref. [4], the Fermi surface of Li-doped system is reported. This represents an important experimental evidence: its comparison with first-principles theoretical prediction will represent a strong fingerprint of the realization of the $\sqrt{3}$, superconducting, phase. In Fig. 2 we report Fig. 1E of Ref. [4] superimposed with calculated Fermi surface of the Li/graphene structure in the $\sqrt{3}$ phase. The agreement between the theoretical and experimental Fermi surfaces is excellent, in particular considering the low spectral weight of the interlayer state. The notable agreement in the size of the Fermi surface indicates that the doping level reached in the experiments is exactly what predicted for the $\sqrt{3}$ phase.

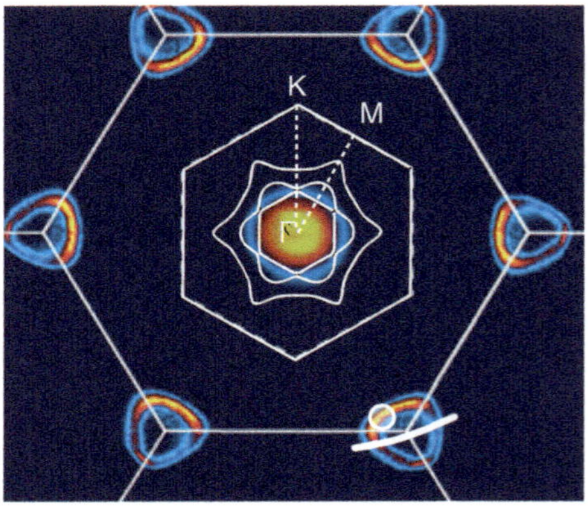

Fig. 2 Experimental (taken from Fig. 1E of Ref. [4]) and theoretical Fermi surface (*white solid line*). The large hexagonal Brillouin zone represents the 1 × 1 graphene BZ, while the inner hexagonal BZ (rotated by 30° with respect to the first one) is the $\sqrt{3}$ BZ of the lithium-doped graphene

We conclude this section summarizing the theoretical results obtained for the superconducting $\sqrt{3}$ phase of lithium-doped graphene: (i) we demonstrate that lithium deposition activate the electron–phonon coupling between B_{2g} out-of-plane modes which induces a sizable splitting of the π bands. (ii) comparison between experimental and theoretical Fermi surfaces at the Γ point reveals an excellent agreement for the interlayer Fermi surface, indicating that the experimental procedure adopted in Ref. [4] indeed produced a $\sqrt{3}$ phase.

3 Metallic Phosphorene: A Novel Superconducting 2D Material

The successful isolation of graphene [1], demonstrated the possibility of use three-dimensional materials in which different layers are brought together by weak van der Waals forces as prototype system for growth of different 2D systems [56]. Indeed, in the last few years, it has been discovered that many different layered bulk materials, like graphite, hexagonal boron nitrite, transition metal dichalcogenides, and iron-chalcogenides [11], posses stable single layer allotropes. These have been obtained experimentally by many different growth methods and represent building blocks of new and possibly useful functional materials. Recently, another member of this family was discovered by means of mechanical exfoliation: the few layer phosphorene[16] was reported and plasma-assisted fabrication [57] allowed the fabrication of the first transistor of single layer black phosphorus [58]. The physical and chemical properties of few layer black phosphorus and its single layer form, phosphorene, are very promising so that they can be considered for nanoelectronic and nanophotonic applications [59]. They combine high electron mobilities like in graphene [58] with the existence of a direct moderate band gap ($\sim$0.3 eV) (in between graphene and transition metal dichalcogenides insulators) which is strongly dependent on the number of layers [60, 61] and strain conditions [62]. Although the semiconducting properties of phosphorene and black phosphorus are so appealing, electron and hole doping represents an important possibility to tune the physical properties and device applications. In principle, doping can be achieved by gating, using an electrolyte gate [63], or by charge-transfer doping as demonstrated in other 2D systems like graphene [64] and MoS_2 [65]. Black phosphorous is completely unexplored in this aspect posing the important question whether it could be functionalized as most of the other 2D materials known. Moreover, charge doping can induce new and unexpected physical properties. For example, superconductivity, one of the most striking phenomena in nature, was recently induced in many 2D systems using suitable substrates [11], electric field [6] and chemical doping [3]. This last possibility is particular appealing considering that moderate doping of 2D semiconductors can result in relatively large superconducting critical temperatures [66]. In this section, we study, by first-principles calculations, the metallization of phosphorene surface by means of metal deposition which moves the Fermi level into the conduction band.

We further explored the effects of the charge doping on the electron–phonon coupling (EPC) predicting a possible superconducting phase of electron and hole doped phosphorene.

3.1 Computational Methods

We performed first-principles DFT calculations within Kohn–Sham local density approximation using norm conserving pseudopotentials, as implemented in the ESPRESSO package [53]. To model a single phosphorene layer, we considered a vacuum space of 15 Å between the periodic replicas. Results are converged using 58 Ry of plane-wave cutoff and a 14 × 14 uniform k-point grid for charge density integration. We relaxed both lattice constants and internal structural parameters in the lowest energy configuration. Phonon and electron–phonon coupling have been calculated within the Linear Response Theory [53] on a 10 × 10 phonon q-grid as referred to the undoped unit cell. Supercells are computed with equivalent grids. The superconducting critical temperature was estimated within Eliashberg theory in form of the Allen–Dynes parameterization [22, 67]:

$$\mathrm{T_C} = f_1 f_2 \frac{\omega_{\log}}{1.2k_B} \exp\left[\frac{-1.04\,(1+\lambda)}{\lambda - \mu^*\,(1+0.62\lambda)}\right] \tag{1}$$

where k_B is the Boltzmann constant and μ^* the Morel–Anderson pseudopotential [21, 68]. λ and $\omega_{\log}$ are, respectively, the electron–phonon coupling parameter and a logarithmic phonon average, defined as

$$\lambda = 2\int \frac{\alpha^2 F(\omega)}{\omega} d\omega \tag{2}$$

$$\omega_{\log} = \exp\left[\frac{2}{\lambda}\int \frac{\ln(\omega)}{\omega}\alpha^2 F(\omega) d\omega\right] \tag{3}$$

by means of the Eliashberg function $\alpha^2 F(\omega)$ [19, 22, 23]. The two prefactors f_1 and f_2 are an important correction to the original McMillan form [28], and make Eq. 1 significantly more accurate. They read:

$$f_1 = \left\{1 + \left[\lambda/2.46\,(1+3.8\mu^*)\right]^{3/2}\right\}^{1/3} \tag{4}$$

$$f_2 = \frac{1 + \lambda^2\left(\omega_2/\omega_{\log} - 1\right)}{\lambda^2 + 3.31\,(1+6.3\mu^*)^2\,\omega_2/\omega_{\log}},$$

where

$$\omega_2 = \left[\frac{2}{\lambda} \int \omega \alpha^2 F(\omega) d\omega \right]^{1/2}. \quad (5)$$

3.2 Rigid Shift Doping

The band structure of a single phosphorene layer is reported as a reference in Fig. 3. Phosphorene is a semiconductor with the top of the valence band (Γ) characterized by a $p_z p_z$ bonding combination while the bottom of the conduction band is of an anti-bonding nature. The undoped system is insulating with a direct band gap at Γ of 0.5 eV in LDA (Fig. 3).

We studied the EPC in phosphorene monolayer and give a possible experimental way to dope the covalent $p_z p_z$ anti-bonding states, making the system metallic. The EPC and its variation with charge doping was studied as a function of the Fermi energy from negative (hole doping) to positive values (electron doping), by a rigid shifting of the Fermi energy considering the phonon frequencies calculated in the undoped (insulating) system. The behavior of the critical temperature as a function

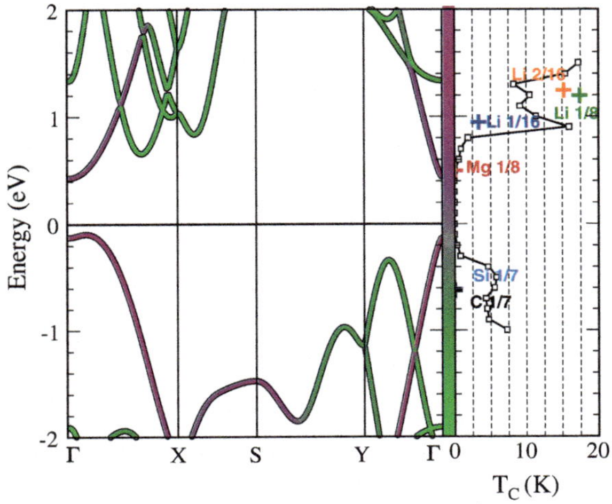

Fig. 3 *Left panel* Kohn–Sham LDA band structure of undoped (insulating) phosphorene. The color-scale (*green* to *violet*) is proportional to the P-p_z character of the KS-state. *Right panel* estimation of the superconducting critical temperature (see text) as a function of the Fermi level position. The zero of the energy represents the valence band maximum of the undoped phosphorene. Points and labels represent the critical temperature obtained considering the inclusion of dopants (see text)

of doping is reported in the right panel of Fig. 3. A maximum T_C of the order of 15 K can be achieved moving the Fermi level at ≃0.5 eV above the conduction band minimum. Hole doping, on the contrary should be less promising. The estimated relevant superconducting critical temperatures upon electron doping deserves further investigations, in particular considering a more detailed and realistic system including dopant atoms. In fact, a rigid shift doping picture does not account either for the local distortion that real doping will induce or, most important, for the screening of phonons when the system becomes metallic [69], and therefore does not consider the possibility of a dynamical instability induced by the doping.

3.3 Realistic Doping by Intercalation or Adsorption

To achieve a more reliable estimation of the superconducting properties of doped phosphorene, we have extended our calculations by simulating real doping by foreign atoms: Si, C, Mg, Li. In particular, lithium doping was simulated at different concentration in order to assess the effect of coverage on the final superconducting critical temperature.

We have used a simulation supercell containing 8 P atoms/unit cell and one substitutional (for hole dopants, Si and C) or adsorbed (for electron dopants, Mg, and Li). While to study lower concentration (1/16) of lithium doping and a different lithium absorption geometry (one lithium atom on both sides of phosphorene layer) we have used a doubled supercell containing 16 P atoms. A sketch of the simulation cells and dopings is shown in the upper panel of Fig. 4. The same figure also shows the calculated α^2F functions. These are used within Eliashberg theory (Allen–Dynes formula of Eq. 1, assuming a $\mu^* = 0.1$) to estimate the superconducting critical temperature.

In Table 1, we report a summary of the relevant parameters which set the superconducting critical temperature for the different systems we considered. As evident, hole dopants (Si and C), in line with estimation based on rigid band approach, do not induce a superconducting phase at relevant critical temperatures. The density of states at the Fermi energy is rather low and so is the integrated coupling λ. On the contrary, electron doping, in particular lithium doping, gives an appreciable superconducting critical temperature with a predicted $T_C = 17.5$ K at 1/8 of coverage. Reducing doping (1/16), or doping on both sides of phosphorene (2/16), decreases the critical temperature. We include the calculated T_C with the rigid doping estimation in Fig. 3. Some considerations are important: although rigid doping effectively gives a good indication that electron–doping is more effective of hole doping to have higher critical temperatures, the quantitative estimation cannot be reliable. The effect of local interaction between different atoms and phosphorene results in different dynamical and electronic properties which cannot be accounted by simple rigid doping. A clear evidence for this effect is given by the inspection of the spectral structure of the $\alpha^2F(\omega)$ for the different systems (see Fig. 4).

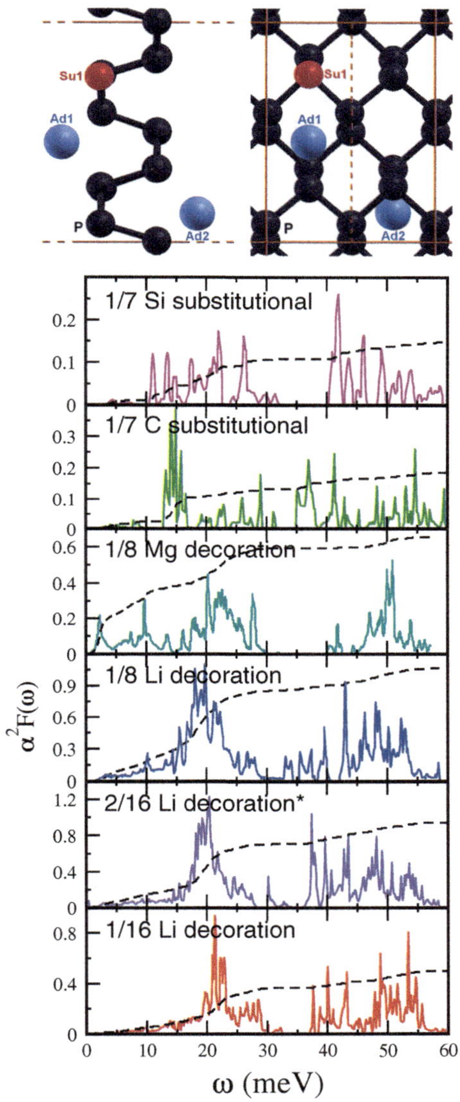

Fig. 4 *Top*: simulation cell of doped phosphorene, side view (*left*) and top view (*right*). The large unit cell (*full orange lines*) contains 16 P atoms, while the smaller one (limited by the *dashed orange line*) contains 8 atoms. Substitutional position (Su1) is shown as a *red dot*. Surface adsorbed positions (Ad1 and Ad2) are shown as *light blue dots*. In the smaller simulation cell we have considered Si and C doping in Su1 and Mg and Li doping in Ad1. In the larger simulation cell a single Li doping in Ad1 and a double Li doping in Ad1 and Ad2. *Bottom*: Eliashberg functions: $\alpha^2F(\omega)$. *Dashed curves* represent $\lambda(\omega) = 2\int_0^{\omega} \frac{\alpha^2F(\omega')}{\omega'}d\omega'$. The label of different systems are consistent with that of Table 1

The origin of the high electron–phonon coupling in lithium-doped systems resides on the increase of the spectral weight in the region of 15–30 (meV), characterized by out-of-plane vibrations [70] of phosphorene. On the contrary, both Si and C are strongly detrimental for the electron–phonon coupling at all the frequencies, while Mg results in a lower coupling due to the higher distance from the phosphorene layer, which reduces the coupling with out-of-plane modes (in line with what observed in graphene). The overall coupling strength of the Mg system is actually high ($\lambda = 0.65$)

Table 1 Couplings and critical temperatures of doped phosphorene systems. λ is the electron–phonon coupling (Eq. 2), ω_{rmlog} the logarithmic averaged (Eq. 3) characteristic frequency (in meV) of the Eliashberg function [22]. T_C is the critical temperature (in K) computed with Eq. 1 and assuming $\mu^* = 0.1$. $N(E_F)$ is the density of states at the Fermi lever (in states/eV/spin). Si and C dopings are substitutional, dopings on a 8 atoms supercell. Mg and Li dopings are decorations of the phosphorene surface. 1/8 dopings are performed on a phosphorene supercell containing 8 P atoms. While 1/16 and 2/16 are performed on a supercell with 16 P atoms. The 2/16 system is decoration is done on both faces of the phosphorene monolayer

	λ	ω_{rmlog}	T_C	$N(E_F)$
1/7 Si	0.14	12.0	0.0	0.35
1/7 C	0.20	20.3	0.0	0.5
1/8 Mg	0.65	8.8	3.2	0.95
1/8 Li	1.06	17.5	17.4	2.12
2/16 Li*	1.00	15.1	13.8	3.78
1/16 Li	0.50	21.7	3.31	1.65

however, as seen in Fig. 4 this is due to an incipient phononic instability (low energy peak in the $\alpha^2 F$ function), however this peak pushes down the logarithmic average of the same function, and with this T_C [22, 67].

4 Conclusions

We have presented first-principles DFT calculations of the electron–phonon coupling properties of doped 2D systems, graphene and phosphorene, underling the main physical mechanisms which sets the superconducting critical temperature. In graphene, the presence of an interlayer band at the Fermi energy is responsible for the increased electron–phonon coupling as confirmed in recent experiments [4]. The comparison between experimental and theoretical Fermi surfaces, indeed confirmed that the superconducting phase is characteristic of the $\sqrt{3}$ surface reconstruction. On the contrary, lithium induces a metallic phase in semiconducting phosphorene. Filling the covalent antibonding p_z states of phosphorous and increasing the density of states at the Fermi energy, lithium makes phosphorene superconducting with a predicted $T_C = 17.5$ K.

We underline that both theoretical predictions were experimentally confirmed: graphene was found superconducting with critical temperature of the order of the predicted one [4], while the possibility to effectively dope phosphorene into a metallic phase was recently demonstrated by ARPES experiments [70].

Acknowledgements G.P. acknowledges support by the Supercomputing center Cineca (Bologna, Italy) through ISCRA projects.

References

1. Novoselov, K.S., Geim, A.K., Morozov, S.V., Jiang, D., Zhang, Y., Dubonos, S.V., Grigorieva, I.V., Firsov, A.A.: Science **306**, 666 (2004)
2. Castro Neto, A.H., Guinea, F., Peres, N.M.R., Novoselov, K.S., Geim, A.K.: Rev. Mod. Phys. **81**, 109 (2009)
3. Profeta, G., Calandra, M., Mauri, F.: Nat. Phys. **8**, 131 (2012)
4. Ludbrook, B.M., Levy, G., Nigge, P., Zonno, M., Schneider, M., Dvorak, D.J., Veenstra, C.N., Zhdanovich, S., Wong, D., Dosanjh, P., Straer, C., Sthr, A., Forti, S., Ast, C.R., Starke, U., Damascelli, A.: Proc. Nat. Acad. Sci. **112**, 11795 (2015). http://www.pnas.org/content/112/38/11795.full.pdf
5. Ye, J.T., Inoue, S., Kobayashi, K., Kasahara, Y., Yuan, H.T., Shimotani, H., Iwasa, Y.: Nat. Mater. **9**, 125 (2010)
6. Ye, J.T., Zhang, Y.J., Akashi, R., Bahramy, M.S., Arita, R., Iwasa, Y.: Science **338**, 1193 (2012)
7. Jo, S., Costanzo, D., Berger, H., Morpurgo, A.F.: Nano Lett. **15**, 1197 (2015), pMID: 25607653. doi:10.1021/nl504314c
8. Shi, W., Ye, J., Zhang, Y., Suzuki, R., Yoshida, M., Miyazaki, J., Inoue, N., Saito, Y., Iwasa, Y.: Sci. Rep. **5**, 12534 (2015)
9. Lu, J.M., Zheliuk, O., Leermakers, I., Yuan, N.F.Q., Zeitler, U., Law, K.T., Ye, J.T.: Science **350**, 1353 (2015). http://science.sciencemag.org/content/350/6266/1353.full.pdf
10. Zhang, T., Cheng, P., Li, W.-J., Sun, Y.-J., Wang, G., Zhu, X.-G., He, K., Wang, L., Ma, X., Chen, X., Wang, Y., Liu, Y., Lin, H.-Q., Jia, J.-F., Xue, Q.-K.: Nat. Phys. **6**, 104 (2010)
11. Qing-Yan, W., Zhi, L., Wen-Hao, Z., Zuo-Cheng, Z., Jin-Song, Z., Wei, L., Hao, D., Yun-Bo, O., Peng, D., Kai, C., Jing, W., Can-Li, S., Ke, H., Jin-Feng, J., Shuai-Hua, J., Ya-Yu, W., Li-Li, W., Xi, C., Xu-Cun, M., Qi-Kun, X.: Chin. Phys. Lett. **29**, 037402 (2012)
12. Ge, J.-F., Liu, Z.-L., Liu, C., Gao, C.-L., Qian, D., Xue, Q.-K., Liu, Y., Jia, J.-F.: Nat. Mater. **14**, 285 (2015)
13. Tao, L., Cinquanta, E., Chiappe, D., Grazianetti, C., Fanciulli, M., Dubey, M., Molle, A., Akinwande, D.: Nat. Nano **10**, 227 (2015)
14. Dvila, M.E., Xian, L., Cahangirov, S., Rubio, A., Lay, G.L.: New J. Phys. **16**, 095002 (2014)
15. Zhu, F.-F., Chen, W.-J., Xu, Y., Gao, C.-L., Guan, D.-D., Liu, C.-H., Qian, D., Zhang, S.-C., Jia, J.-F.: Nat. Mater. **14**, 1020 (2015)
16. Li, L., Yu, Y., Ye, G.J., Ge, Q., Ou, X., Wu, H., Feng, D., Chen, X.H., Zhang, Y.: Nat. Nano **9**, 372 (2014)
17. Mannix, A.J., Zhou, X.-F., Kiraly, B., Wood, J.D., Alducin, D., Myers, B.D., Liu, X., Fisher, B.L., Santiago, U., Guest, J.R., Yacaman, M.J., Ponce, A., Oganov, A.R., Hersam, M.C., Guisinger, N.P.: Science **350**, 1513 (2015). http://science.sciencemag.org/content/350/6267/1513.full.pdf
18. Kohn, W., Sham, L.J.: Phys. Rev. **140**, A1133 (1965)
19. Eliashberg, G.: J. Exptl. Theor. Phys. (U.S.S.R.) 11, 966, : Sov. Phys. JETP **38**, 696 (1960)
20. Oliveira, L.N., Gross, E.K.U., Kohn, W.: Phys. Rev. Lett. **60**, 2430 (1988)
21. Scalapino, D.J., Schrieffer, J.R., Wilkins, J.W.: Phys. Rev. **148**, 263 (1966)
22. Allen, P.B., Mitrović, B.: Solid State Phys. **37** (1960)
23. Carbotte, J.P.: Rev. Modern Phys. **62**, 1027 (1990)
24. Margine, E.R., Giustino, F.: Phys. Rev. B **87**, 024505 (2013)
25. Sanna, A., Pittalis, S., Dewhurst, J.K., Monni, M., Sharma, S., Ummarino, G., Massidda, S., Gross, E.K.U.: Phys. Rev. B **85**, 184514 (2012)
26. Pietronero, L., Strässler, S., Grimaldi, C.: Phys. Rev. B **52**, 10516 (1995)
27. Grimaldi, C., Pietronero, L., Strässler, S.: Phys. Rev. B **52**, 10530 (1995)
28. McMillan, W.L.: Phys. Rev. **167**, 331 (1968)
29. Lüders, M., Marques, M.A.L., Lathiotakis, N.N., Floris, A., Profeta, G., Fast, L., Continenza, A., Massidda, S., Gross, E.K.U.: Phys. Rev. B **72**, 024545 (2005)
30. Essenberger, F., Sanna, A., Linscheid, A., Tandetzky, F., Profeta, G., Cudazzo, P., Gross, E.K.U.: Phys. Rev. B **90**, 214504 (2014)

31. Akashi, R., Arita, R.: Phys. Rev. Lett. **111**, 057006 (2013)
32. Marques, M.A.L., Lüders, M., Lathiotakis, N.N., Profeta, G., Floris, A., Fast, L., Continenza, A., Gross, E.K.U., Massidda, S.: Phys. Rev. B **72**, 024546 (2005)
33. Floris, A., Sanna, A., Massidda, S., Gross, E.K.U.: Phys. Rev. B **75**, 054508 (2007)
34. Floris, A., Profeta, G., Lathiotakis, N.N., Lüders, M., Marques, M.A.L., Franchini, C., Gross, E.K.U., Continenza, A., Massidda, S.: Phys. Rev. Lett. **94**, 037004 (2005)
35. Linscheid, A., Sanna, A., Floris, A., Gross, E.K.U.: Phys. Rev. Lett. **115**, 097002 (2015)
36. Sanna, A., Profeta, G., Floris, A., Marini, A., Gross, E.K.U., Massidda, S.: Phys. Rev. B **75**, 020511 (2007)
37. Flores-Livas, J.A., Sanna, A.: Phys. Rev. B **91**, 054508 (2015)
38. Flores-Livas, J.A., Sanna, A., Gross, E.K.U.: Eur. Phys. J. B **89**, 63 (2016)
39. Flores-Livas, J.A., Amsler, M., Heil, C., Sanna, A., Boeri, L., Profeta, G., Wolverton, C., Goedecker, S., Gross, E.K.U.: Phys. Rev. B **93**, 020508 (2016)
40. Cudazzo, P., Profeta, G., Sanna, A., Floris, A., Continenza, A., Massidda, S., Gross, E.K.U.: Phys. Rev. Lett. **100**, 257001 (2008)
41. Cudazzo, P., Profeta, G., Sanna, A., Floris, A., Continenza, A., Massidda, S., Gross, E.K.U.: Phys. Rev. B **81**, 134505 (2010)
42. Cudazzo, P., Profeta, G., Sanna, A., Floris, A., Continenza, A., Massidda, S., Gross, E.K.U.: Phys. Rev. B **81**, 134506 (2010)
43. Bersier, C., Floris, A., Sanna, A., Profeta, G., Continenza, A., Gross, E.K.U., Massidda, S.: Phys. Rev. B **79**, 104503 (2009)
44. Sanna, A., Franchini, C., Floris, A., Profeta, G., Lathiotakis, N.N., Lüders, M., Marques, M.A.L., Gross, E.K.U., Continenza, A., Massidda, S.: Phys. Rev. B **73**, 144512 (2006)
45. Profeta, G., Franchini, C., Lathiotakis, N.N., Floris, A., Sanna, A., Marques, M.A.L., Lüders, M., Massidda, S., Gross, E.K.U., Continenza, A.: Phys. Rev. Lett. **96**, 047003 (2006)
46. Drozdov, A.P., Eremets, M.I., Troyan, I.A., Ksenofontov, V., Shylin, S.I.: Nature **000** (2015)
47. Duan, D., Liu, Y., Tian, F., Li, D., Huang, X., Zhao, Z., Yu, H., Liu, B., Tian, W., Cui, T.: Sci. Rep. **4** (2014). doi:10.1038/srep06968
48. Heersche, H.B., Jarillo-Herrero, P., Oostinga, J.B., Vandersypen, L.M.K., Morpurgo, A.F.: Nature **446**, 56 (2007)
49. Pathak, S., Shenoy, V.B., Baskaran, G.: Phys. Rev. B **81**, 085431 (2010)
50. Calandra, M., Mauri, F.: Phys. Rev. B **74**, 094507 (2006)
51. Boeri, L., Bachelet, G.B., Giantomassi, M., Andersen, O.K.: Phys. Rev. B **76**, 064510 (2007)
52. Perdew, J.P., Zunger, A.: Phys. Rev. B **23**, 5048 (1981)
53. Giannozzi, P., Baroni, S., Bonini, N., Calandra, M., Car, R., Cavazzoni, C., Ceresoli, D., Chiarotti, G.L., Cococcioni, M., Dabo, I., Corso, A.D., de Gironcoli, S., Fabris, S., Fratesi, G., Gebauer, R., Gerstmann, U., Gougoussis, C., Kokalj, A., Lazzeri, M., Martin-Samos, L., Marzari, N., Mauri, F., Mazzarello, R., Paolini, S., Pasquarello, A., Paulatto, L., Sbraccia, C., Scandolo, S., Sclauzero, G., Seitsonen, A.P., Smogunov, A., Umari, P., Wentzcovitch, R.M.: J. Phys. Condens. Matter **21**, 395502 (2009)
54. Yang, S.-L., Sobota, J.A., Howard, C.A., Pickard, C.J., Hashimoto, M., Lu, D.H., Mo, S.-K., Kirchmann, P.S., Shen, Z.-X.: Nat. Commun. **5** (2014)
55. Sugawara, K., Sato, T., Takahashi, T.: Nat. Phys. **5**, 40 (2009)
56. Geim, A.K., Grigorieva, I.V.: Nature **499**, 419 (2013)
57. Lu, W., Nan, H., Hong, J., Chen, Y., Zhu, C., Liang, Z., Ma, X., Ni, Z., Jin, C., Zhang, Z.: Nano Res. **7**, 853 (2014)
58. Liu, H., Neal, A.T., Zhu, Z., Luo, Z., Xu, X., Tománek, D., Ye, P.D.: ACS Nano **8**, 4033 (2014), pMID: 24655084
59. Youngblood, N., Chen, C., Koester, S.J., Li, M.: Nat. Photon **9**, 247 (2015)
60. Tran, V., Soklaski, R., Liang, Y., Yang, L.: Phys. Rev. B **89**, 235319 (2014)
61. Cai, Y., Zhang, G., Zhang, Y.-W.: Sci. Rep. **4**, 6677 EP (2014)
62. Fei, R., Yang, L.: Nano Lett. **14**, 2884 (2014), pMID: 24779386
63. Das, A., Pisana, S., Chakraborty, B., Piscanec, S., Saha, S.K., Waghmare, U.V., Novoselov, K.S., Krishnamurthy, H.R., Geim, A.K., Ferrari, A.C., Sood, A.K.: Nat. Nano **3**, 210 (2008)

64. Fedorov, A.V., Verbitskiy, N.I., Haberer, D., Struzzi, C., Petaccia, L., Usachov, D., Vilkov, O.Y., Vyalikh, D.V., Fink, J., Knupfer, M., Büchner, B., Grüneis, A.: Nat. Commun. **5** (2014)
65. Miwa, J.A., Ulstrup, S., Sørensen, S.G., Dendzik, M., Antonija, G., Bianchi, M., Lauritsen, J.V., Hofmann, P.: Phys. Rev. Lett. **114**, 046802 (2015)
66. Calandra, M., Zoccante, P., Mauri, F.: Phys. Rev. Lett. **114**, 077001 (2015)
67. Allen, P.B., Dynes, R.C.: Phys. Rev. B **12**, 905 (1975)
68. Morel, P., Anderson, P.W.: Phys. Rev. **125**, 1263 (1962)
69. Baroni, S., de Gironcoli, S., Dal Corso, A., Giannozzi, P.: Rev. Mod. Phys. **73**, 515 (2001)
70. Sanna, A., Fedorov, A.V., Verbitskiy, N., Fink, J., Krellner, C., Petaccia, L., Chikina, A., Usachov, D., Grüneis, A., Profeta, G.: submitted (2016)

Elastic Properties and Electron–Phonon Coupling of Graphene/Metal Interfaces Probed by Phonon Dispersion

M. Alfano, C. Lamuta, G. Chiarello and A. Politano

Abstract High-resolution electron energy loss spectroscopy is a suitable tool for investigating phonons in graphene, due to its exceptional energy resolution in both the energy and momentum domains. In this chapter, we show that the experimental phonon dispersion of graphene can be used to estimate elastic properties and electron–phonon coupling. Novel coupling mechanisms of Dirac cone electrons in graphene with out-of-plane optical phonons of the graphene lattice, activated only whenever the graphene sheet is supported by a solid substrate, are also discussed.

1 Introduction

Most of the amazing properties of graphene arise from the linear energy dispersion of the conduction and valence bands near the Fermi level [17, 25]. Massless charge carriers in the Dirac cone behave as Dirac fermions with a group velocity ≈1/300 of the speed of light [52].

The superb control enabled by epitaxial growth of graphene facilitates the spectroscopic characterization of the system [9, 87]. The preparation of highly ordered graphene samples could be extended up to the millimeter scale when graphene is epitaxially grown on transition-metal substrates [54] by chemical vapor deposition (CVD).

The investigation of the interaction strength between the graphene layer and the metallic substrate is important in order to discern between physisorption and

M. Alfano · G. Chiarello · A. Politano (✉)
Department of Physics, University of Calabria, Ponte P. Bucci,
Cubo 31C, 87036 Rende, Italy
e-mail: antonio.politano@iit.it

C. Lamuta
Department of Mechanical, Energy and Management Engineering
University of Calabria, Ponte P. Bucci, Cubo 44C, 87036 Rende, Italy

A. Politano
Fondazione Istituto Italiano di Tecnologia, Graphene Labs,
via Morego 30, 16163 Genoa, Italy

V. Morandi and L. Ottaviano (eds.), *GraphITA*, Carbon Nanostructures,
DOI 10.1007/978-3-319-58134-7_4

chemisorption of graphene and, thus, to appraise the quality of the contacts between metallic electrodes and graphene devices.

Monolayer graphene (MLG) has been grown on different transition-metal substrates: [24, 27, 40, 51, 60, 61, 66, 68, 77]. Among the above-mentioned graphene systems, three general classes may be distinguished.

First, for the class of Ni and Co substrates, the mismatch in the lattice parameter is negligible, thus the MLG unit cell may be directly matched with the substrate unit cell by slightly quenching or stretching the bonds between carbon atoms of the graphene lattice. In this case, a strong hybridization between the substrate d-bands and the π-bands of MLG occurs [30]. In addition, the distance between the graphene sheet and the substrate is reduced (2.1 Å for Ni [29] and 1.5–2.2 Å for Co [27]).

Whenever the mismatch in the lattice parameter approaches 5–10%, a Moiré pattern appears. In this case, the graphene sheet may be weakly (Pt, Ir) or strongly bound (Ru, Re, Pd, Rh) to the substrate. The strong interaction occurring for graphene on [51, 83] causes a strong corrugation of the graphene sheet.

Vibrations of the graphene lattice are characterized by three types of phonons. Modes classified with "T" are shear in-plane phonon excitations; "L" modes are longitudinal in-plane vibrations; while "Z" indicates vibrations out of the plane of the layer—the so-called flexural phonons. In turn, they can be acoustic (A) or optical (O). Thus, graphene and graphite have six distinct phonons: TA, TO, LA, LO, ZA, and ZO.

The dynamics of atoms at surfaces plays an important role in many chemical and physical processes. In particular, lattice vibrations can provide essential information on many physical properties, such as thermal expansion [79], heat capacity [45], magnetic forces [10], and thermal conductivity [19]. Phonon modes in graphene/metal interfaces can host different intriguing phenomena, such as electron–phonon coupling with the emergence of Kohn anomalies not activated in free-standing graphene [63], the modification of phonon dispersion [3], and different selection rules in phonon excitation [23].

Both the dispersion relation of phonon modes of the graphene overlayer and their coupling mechanism with Dirac cone electrons are affected by the presence of the underlying metal substrates, as we will show in this chapter.

2 Elastic Parameters Evaluated from Phonon Dispersion

Understanding the elastic properties is obligatory in order to tailor graphene's mechanical properties. The intrinsic strength of graphene, which is higher with respect to other materials [90], allows its use for several applications, such as actuators [55] and nano-electromechanical devices [74, 86] and, moreover, as carbon-fiber reinforcement in polymeric nanocomposites [4].

Atomic force microscopy (AFM) is a suitable tool for investigating elastic properties of graphene [43, 69, 80]. The main benefits of AFM for such experiments are the lateral resolution and the imaging capability at the nanoscale. However, the tip shape and the contact geometry between tip and sample, which are unknown, represent a remarkable hurdle to extensive and precise measurements.

In Refs. [62, 64], an alternative in situ methodology for calculating the average elastic properties (Young's modulus and the Poisson's ratio) of epitaxial graphene, based on the investigation of the phonon dispersion by high-resolution electron energy loss spectroscopy (HREELS), has been presented. Therein, the average elastic properties (Young's modulus and the Poisson's ratio) in graphene epitaxially grown on Ru(0001), Pt(111), Ir(111), Ni(111), BC3/NbB2(0001) and graphite have been investigated.

Firstly, the bending rigidity τ, an important parameter for mechanical properties of graphene membranes, has been evaluated from the quadratic dispersion of the ZA phonon:

$$\omega_{ZA}(q_{||}) = \sqrt{\frac{\tau}{\rho_{2D}}}|\vec{q}_{||}|^2, \quad (1)$$

where $\rho_{2D} = 4m_C/(3\sqrt{3}a^2)$ is the two-dimensional (2D) mass density (mC is the atomic mass of carbon atoms and a is the in-plane lattice parameter). Experimental data show a value of (1.30 ± 0.15) eV for bending rigidity τ [2].

Sound velocities have been extracted from the experimental slope of the acoustic branches in the low-q|| limit, for which TA and LA phonons along the $\Gamma - \bar{K}$ and $\Gamma - \bar{M}$ directions coincide. In particular, vL, longitudinal sound velocity, and vT, transverse sound velocity, are defined as $v_L = \frac{d\omega_{LA}}{dq}$ and $v_T = \frac{d\omega_{TA}}{dq}$ where ω_{LA} and ω_{TA} are the frequencies of LA and TA phonons, respectively.

For most systems, 14.0 and 22.0 km/s are obtained for the TA and the LA branches, respectively. The only exception is the transverse sound velocity for MLG/Ni(111) (12.4 km/s). The reduced value obtained for MLG/Ni(111) is related to the stretching of the C – C bonds of the graphene layer by 1.48% to form a 1 × 1 structure [3].

These results agree with sound velocities for single-crystalline graphite determined using inelastic x-ray scattering [13], i.e., 14.7 and 22.2 km/s. Previous HREELS investigation in Ref. [53] for the surface phonons of graphite found sound velocities of 14 and 24 km/s for TA and LA modes, respectively.

According to the procedure illustrated in Ref. [71], the in-plane stiffness κ (the 2D analogous of the bulk modulus) and the shear modulus μ of the graphene sheet can be determined from the sound velocities of the TA and LA branches, respectively:

$$v_L = \sqrt{\frac{\kappa + \mu}{\rho_{2D}}}$$
$$v_T = \sqrt{\frac{\mu}{\rho_{2D}}} \tag{2}$$

Thus, values of κ and μ of 211 and 144 N/m are valued, respectively. Again, the only exception is represented by the case of MLG/Ni(111), for which the values of κ and μ are 244 and 114 N/m, respectively. It is useful to note that graphene is a true 2D material, and, thus, its elastic behavior is properly described by 2D properties with units of force/length.

On the other hand, the 2D shear and bulk moduli are also defined as a function of the Poisson's ratio ν and of the 2D Young's modulus E2D

$$\kappa = \frac{E^{2D}}{2(1-\nu)}$$
$$\mu = \frac{E^{2D}}{2(1+\nu)} \tag{3}$$

Therefore, from κ and μ it is possible to estimate the Poisson's ratio, i.e., the ratio of transverse contraction strain to longitudinal extension strain in the direction of the stretching force:

$$\upsilon = \frac{\frac{\kappa}{\mu} - 1}{\frac{\kappa}{\mu} + 1} \tag{4}$$

In most cases, Poisson's ratio and the Young's modulus are 0.19 and 342 N/m, respectively. The unique exception is represented by MLG/Ni(111), for which their values are 0.36 and 310 N/m, respectively. For the sake of comparison, the Poisson's ratio for graphite in the basal plane is 0.165 [12, 75] while it is 0.28 in carbon nanotubes [47]. Table 1 shows a comparison with values in literature. Despite the macroscopic size of graphene samples used in HREELS, which usually reduces the tensile strength for the presence of defects [88] and grain boundaries [16, 33], the above parameters well agree with results reported for suspended graphene membranes [43] with diameter of 1.0–1.5 μm.

It is also possible to estimate the Young's modulus E2D, which is a measure of the stiffness of an isotropic elastic material. It is defined as the ratio of the uniaxial stress over the uniaxial strain.

As reported in Table 2, many theoretical works found Young's moduli ranging from 307 to 356 N/m. The obtained value of E2D for most systems, i.e. 342 N/m, agrees well with most theoretical results (Table 2), a part from calculations in Ref. [92] (underestimated E2D). A lower value of E2D is found for MLG/Ni(111), i.e., 310 N/m.

Table 1 Poisson's ratio ν, as reported in different experimental and theoretical works

	Poisson's ratio ν
Experimental (HREELS), graphene on Pt(111), Ru(0001), Ir(111), BC3/NbB2 (0001), graphite, Ref. [62]	0.19
Experimental, basal plane of graphite, Refs. [12, 75]	0.165
Experimental (HREELS), MLG/Ni(111), Ref. [62]	0.36
Atomistic Monte Carlo, Ref. [89]	0.12
Tersoff-Brenner potential, Ref. [39]	0.149
Continuum plate theory, Ref. [5]	0.16
Density functional theory, Ref. [11]	0.162
First-principles total-energy calculations, combined to continuum elasticity, Ref. [15]	0.169
Ab initio, Ref. [31]	0.173
Ab initio, Ref. [56]	0.178
Density functional theory, Ref. [28]	0.18
Ab initio, Ref. [46]	0.186
Ab initio, Ref. [85]	0.19
Valence force model, Ref. [57]	0.20
Local-density approximation, Ref. [82]	0.20
Cellular material mechanics theory, Ref. [73]	0.21
Molecular dynamics, Ref. [34]	0.22
Molecular dynamics, Ref. [35]	0.22
Empirical force-constant calculations, Ref. [48]	0.227
Brenner's potential, Ref. [28]	0.27
continuum elasticity theory and tight-binding atomistic simulations, Ref. [14]	0.31
Ab initio, Ref. [91]	0.32
Molecular dynamics, Ref. [36]	0.32
Brenner's potential, Ref. [6]	0.397
Multiple component correlation model, Ref. [84]	0.4
Molecular dynamics, Ref. [93]	0.45

3 Electron–Phonon Coupling (EPC) and Kohn Anomalies

The interaction between electrons and phonons is a particularly interesting field in condensed-matter physics. In particular, understanding physical phenomena that govern lattice dynamics and the EPC in graphene might afford essential information on its novel and unusual properties. EPC in graphene [3] deserves a particular attention, since lattice dynamics plays a key role on thermal properties of graphene. Furthermore, the EPC limits the ballistic electronic transport [42] in graphene-based electronic devices. Therefore, the EPC can be considered as the main impediment for ballistic transport. Atomic vibrations might be screened by electrons and, moreover, screening can change rapidly for vibrations associated with high-symmetry points of

Table 2 2D Young's modulus E2D, expressed in N/m, as reported in different experimental and theoretical works

	Young's modulus E2D (N/m)
Experimental (HREELS), graphene on Pt(111), Ru(0001), Ir(111), BC3/NbB2(0001), graphite Ref. [62]	342
Experimental (HREELS), MLG/Ni(111), Ref. [62]	310
Experimental (AFM) on graphene/copper foils, Ref. [44]	339 ± 17
Experimental (AFM) on graphene membranes, Ref. [43]	340 ± 50
Experimental (AFM) on graphene membranes, Ref. [20]	350 ± 50
Tersoff-Brenner potential, Ref. [92]	235
Energetic model, Ref. [67]	307
continuum elasticity theory and tight-binding atomistic simulations, Ref. [14]	312
Density functional theory, Ref. [28]	330
Brenner's potential, Ref. [6]	336
First-principles total-energy calculations, combined to continuum elasticity, Ref. [15]	344
Tersoff-Brenner potential, Ref. [39]	345
Ab initio, Ref. [46]	350
Atomistic Monte Carlo, Ref. [89]	353
Density functional theory, Ref. [11]	356
Empirical force-constant calculations, Ref. [48]	384
Experimental (AFM) on graphene/copper foils, Ref. [70]	55

the Brillouin zone (BZ). This phenomenon leads to an unusual behavior of the phonon dispersion around such points, which is called Kohn anomaly [38]. Kinks in phonon dispersion indicate the occurrence of Kohn anomalies [3, 7, 18, 22, 37, 58, 59, 65, 72, 81], which are a manifestation of the coupling between electrons and phonons, and their existence is completely determined by the shape of the Fermi surface.

In graphite [58], Kohn anomalies are realized as linear cusps in the dispersion for the highest optical phonon branches at Γ (LO phonon) and at K (TO phonon). Their existence is associated to the dispersion of the π bands around the high-symmetry point K.

Several theoretical studies predicted the existence of Kohn anomalies [21, 22, 41, 49, 50, 81], Hwang and Das Sarma (32) for graphene for Γ-LO and K-TO phonons. The form and position of cusps is determined by the dispersion of the Dirac cone and the Fermi level $\mu \approx 0$. These two in-plane phonon branches have been for a long time commonly accepted to be the only ones with a significant EPC in graphene [16]. Experimentally, HREELS has been used to probe Kohn anomalies in MLG/Ir(111) [26] and MLG/Pt(111) [65].

Figure 1 shows the dispersion relations of the LO phonon around Γ (panel a of Fig. 1) and of the TO phonon nearby K (Fig. 1b) for the case of MLG/Ir(111). It

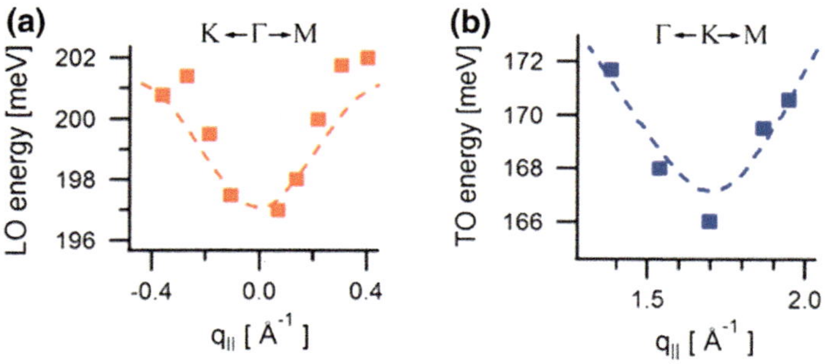

Fig. 1 Dispersion of **a** the LO phonon in the nearness of Γ and of **b** the TO phonon close to K. *Dashed lines* show Local-density approximation calculations for graphene on Ir(111). Adapted from Ref. [26]

can be concluded that Kohn anomaly of the highest optical phonon branch at Γ persists in this system.

Theoretical works also indicates that Dirac fermions of graphene in the presence of Coulomb interactions show power-law behavior, depending on the coupling strength β [22]. The existence of power-law Kohn anomalies has been experimentally corroborated for the case of MLG/Pt(111) [63]. HREELS experiments [63] have also revealed a novel coupling mechanism of electrons with out-of-plane optical phonons. In free-standing graphene, the EPC for out-of-plane phonons is strongly constrained by the presence of mirror symmetry with respect to the horizontal plane, which forbids a first-order coupling to electrons. Nevertheless, when graphene is supported by a substrate, a first-order EPC becomes allowed.

Figure 2 shows the phonon dispersion of the LO and ZO phonons in MLG/Pt (111) for small momenta. Both phonons exhibit a cusp at the identical momentum $q \sim 0.13$ Å $- 1$, arising from Kohn anomalies at $q = 2kF$. The Fermi wave-vector $kF = EF/vF$ can be estimated from angle-resolved photoemission spectroscopy (ARPES) measurements of MLG/Pt(111) [78]. The reported values of the Fermi energy and Fermi velocity are $EF \approx 0.30 \pm 0.15$ eV and $vF \approx 6$ eV · Å. Hence, $2kF \approx 0.10 \pm 0.05$ Å $- 1$, in accordance to the position of the cusp found in the dispersion relation of LO and ZO modes in Fig. 15 [63].

In Ref. [63], a dimensionless EPC λi has been introduced for the optical phonon i (i = LO, TO, ZO)

$$\lambda_i \frac{F_i^2 A_c}{2M\omega_i v_F^2} \tag{6}$$

Here Fi is the electron–phonon coupling as defined in Ref. [8], Ac is the unit cell area, and ωi is the energy of the i phonon mode.

The EPC induces a phonon self-energy Πi(q) that corrects the dispersion according to

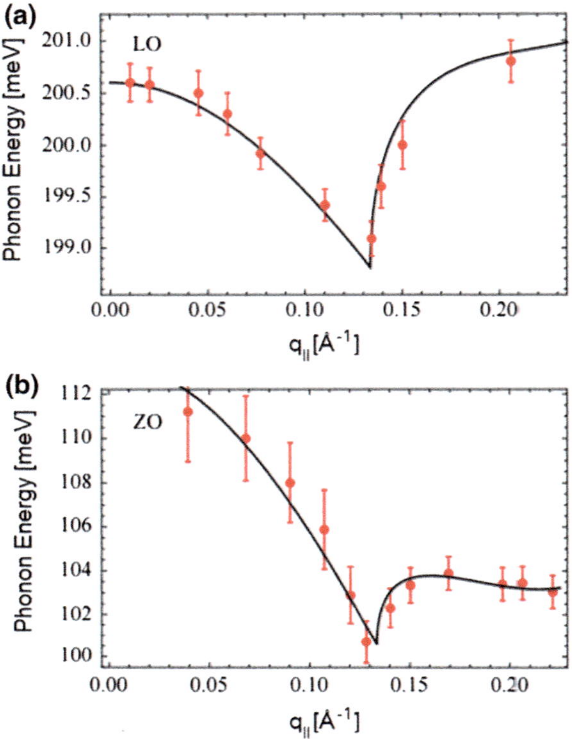

Fig. 2 Phonon dispersions in the nearness of the Γ point, showing cusps around qc ~ 0.13 Å − 1. **a** Dispersion of the LO phonon. A fit to Eq. [7] (full line) gives an EPC λLO ~ 0.029, and parameters ELO(q = 0) = 200.6 meV, aLO = 110 meV · Å2, bLO = 558 meV · Å4. **b** Dispersion of the ZO phonon. A fit to Eq. [7] (full line) gives an EPC λZO ~ 0.087 and EZO (q = 0) = 102.3 meV, aZO = 810 meV · Å2, bZO = 5243 meV · Å4. Adapted from Ref. [63]

$$\omega_{R,i} = \omega_i^0 + a_i q^2 + b_i q^4 + \frac{\lambda_i}{2}\Pi_i(q/k_F) \tag{7}$$

In the static approximation, the self-energies for the different phonon branches in terms of the dimensionless variable x = q/kF are given by

$$\begin{aligned} \Pi_{\mathrm{LO}}(x) &= \frac{g_{s},\nu\mu}{4\pi}\left(\sqrt{1-\frac{4}{x^2}}+\frac{x}{2}\arccos(2/x)\right)\theta(2-x), \\ \Pi_{\mathrm{TO}}(x) &= 0. \\ \Pi_{\mathrm{ZO}}(x) &= \frac{g_{s,\nu}\mu}{4\pi}\left[2+x\,\arccos(2/x)\theta(2-x)\right]. \end{aligned} \tag{8}$$

These equations were used to fit the experimental phonon dispersion showing cusps to estimate the different values of the EPC for both LO and ZO modes. A best-fit gives a value for the Fermi energy EF = $v_F q_c/2 \approx 0.401$ eV, which corresponds to a cusp momentum of q_c = 0.133 Å − 1.

The LO EPC has been determined to be λLO ~ 0.029, in excellent agreement with its estimation by means of Raman spectroscopy (0.027–0.034) [8]. For the ZO mode, the fit procedure gives λZO ~ 0.087, an even greater value.

The considerable magnitude λZO implies that its effects could be resolved in future ARPES experiments. The finite value of λZO is expected also to influence transport properties at high bias voltages.

On the other hand, the disappearance of the Kohn anomalies is the most evident peculiarity of phonon dispersion in MLG/Ni(111). In this system, the dispersion of the highest optical phonon branches is nearly flat both at Γ [1, 76]. Therefore, it can be concluded that the interaction with the substrate leads to the complete suppression of Kohn anomalies, as a consequence of the strong hybridization of the graphene π-bands with Ni d-bands [30].

4 Conclusions and Outlook

Herein, some aspects related to phonon excitations in epitaxial graphene systems have been shown and discussed. We have evidenced that the experimental phonon dispersion could give information on the elastic properties of the graphene sheet. Moreover, the influence of the substrate on the out-of-plane optical phonons and their coupling with Dirac cone electrons have been highlighted. Nevertheless, numerous issues are still open. As an example, the influence of charge doping on the ZO Kohn anomaly is still unclear. The effects of plasmon–phonon coupling on plasmonic devices are also hitherto unexplored. Therefore, the study of phonon excitations in epitaxial graphene is expected to continue engaging researchers in next years.

References

1. Aizawa, T., Souda, R., Ishizawa, Y., Hirano, H., Yamada, T., Tanaka, K.-I., Oshima, C.: Phonon dispersion in monolayer graphite formed on Ni(111) and Ni(001). Surf. Sci. **237**, 194–202 (1990)
2. AL Taleb, A., Yu, H.K., Anemone, G., Farías, D., Wodtke, A.M.: Helium diffraction and acoustic phonons of graphene grown on copper foil. Carbon **95**, 731–737 (2015)
3. Allard, A., Wirtz, L.: Graphene on metallic substrates: suppression of the Kohn anomalies in the phonon dispersion. Nano Lett. **10**, 4335–4340 (2010)
4. Alzari, V., Nuvoli, D., Sanna, R., Scognamillo, S., Piccinini, M., Kenny, J.M., Malucelli, G., Mariani, A.: In situ production of high filler content graphene-based polymer nanocomposites by reactive processing. J. Mater. Chem. **21**, 16544–16549 (2011)
5. Arghavan, S., Singh, A.V.: Free vibration of single layer graphene sheets: lattice structure versus continuum plate theories. J. Nanotechnol. Eng. Med. **2**, 031005–031006 (2011)
6. Arroyo, M., Belytschko, T.: Finite crystal elasticity of carbon nanotubes based on the exponential Cauchy-Born rule. Phys. Rev. B **69**, 115415 (2004)

7. Aynajian, P., Keller, T., Boeri, L., Shapiro, S.M., Habicht, K., Keimer, B.: Energy gaps and kohn anomalies in elemental superconductors. Science **319**, 1509–1512 (2008)
8. Basko, D.M., Piscanec, S., Ferrari, A.C.: Electron-electron interactions and doping dependence of the two-phonon Raman intensity in graphene. Phys. Rev. B **80**, 165413 (2009)
9. Batzill, M.: The surface science of graphene: Metal interfaces, CVD synthesis, nanoribbons, chemical modifications, and defects. Surf. Sci. Rep. **67**, 83–115 (2012)
10. Benedek, G., Hulpke, E., Steinhogl, W.: Probing the magnetic forces in fcc-Fe(001) films by means of surface phonon spectroscopy. Phys. Rev. Lett. **87**, 027201 (2001)
11. Bera, S., Arnold, A., Evers, F., Narayanan, R., Wölfle, P.: Elastic properties of graphene flakes: boundary effects and lattice vibrations. Phys. Rev. B **82**, 195445 (2010)
12. Blakslee, O.L., Proctor, D.G., Seldin, E.J., Spence, G.B., Weng, T.: Elastic constants of compression-annealed pyrolytic graphite. J. Appl. Phys. **41**, 3373–3382 (1970)
13. Bosak, A., Krisch, M., Mohr, M., Maultzsch, J., Thomsen, C.: Elasticity of single-crystalline graphite: inelastic x-ray scattering study. Phys. Rev. B **75**, 153408 (2007)
14. Cadelano, E., Palla, P.L., Giordano, S., Colombo, L.: Nonlinear elasticity of monolayer graphene. Phys. Rev. Lett. **102**, 235502 (2009)
15. Cadelano, E., Palla, P.L., Giordano, S., Colombo, L.: Elastic properties of hydrogenated graphene. Phys. Rev. B **82**, 235414 (2010)
16. Cao, A., Yuan, Y.: Atomistic study on the strength of symmetric tilt grain boundaries in graphene. Appl. Phys. Lett. **100**, 211912 (2012)
17. Castro Neto, A.H., Guinea, F., Peres, N.M.R., Novoselov, K.S., GEIM, A.K.: The electronic properties of graphene. Rev. Mod. Phys. **81**, 109–162 (2009)
18. Caudal, N., Saitta, A.M., Lazzeri, M., Mauri, F.: Kohn anomalies and nonadiabaticity in doped carbon nanotubes. Phys. Rev. B **75**, 115423 (2007)
19. Chis, V., Benedek, G.: Phonon-Induced surface charge density oscillations in quantum wells: a first-principles study of the (2 × 2)-K overlayer on Be(0001). J. Phys. Chem. A **115**, 7242–7248 (2011)
20. Clark, N., Oikonomou, A., Vijayaraghavan, A.: Ultrafast quantitative nanomechanical mapping of suspended graphene. Physica Status Solidi (b) **250**, 2672–2677 (2013)
21. de Juan, F., Fertig, H.A.: Power-law Kohn anomaly in undoped graphene induced by Coulomb interactions. Phys. Rev. B **85**, 085441 (2012)
22. de Juan, F., Fertig, H.A.: Power law kohn anomalies and the excitonic transition in graphene. Solid State Commun. **152**, 1460–1468 (2012)
23. de Juan, F., Politano, A., Chiarello, G., Fertig, H.A.: Symmetries and selection rules in the measurement of the phonon spectrum of graphene and related materials. Carbon **85**, 225–232 (2015)
24. Dedkov, Y.S., Fonin, M., Laubschat, C.: A possible source of spin-polarized electrons: the inert graphene/Ni(111) system. Appl. Phys. Lett. **92**, 052506 (2008)
25. Elias, D.C., Gorbachev, R.V., Mayorov, A.S., Morozov, S.V., Zhukov, A.A., Blake, P., Ponomarenko, L.A., Grigorieva, I.V., Novoselov, K.S., Guinea, F., Geim, A.K.: Dirac cones reshaped by interaction effects in suspended graphene. Nat. Phys. **7**, 701–704 (2011)
26. Endlich, M., Molina-Sánchez, A., Wirtz, L., Kröger, J.: Screening of electron-phonon coupling in graphene on Ir(111). Phys. Rev. B **88**, 205403 (2013)
27. Eom, D., Prezzi, D., Rim, K.T., Zhou, H., Lefenfeld, M., Xiao, S., Nuckolls, C., Hybertsen, M.S., Heinz, T.F., Flynn, G.W.: Structure and electronic properties of graphene nanoislands on Co(0001). Nano Lett. **9**, 2844–2848 (2009)
28. Fair, K.M., Arnold, M.D., Ford, M.J.: Determination of the elastic properties of graphene by indentation and the validity of classical models of indentation. J. Phys.: Condens. Matter **26**, 015307 (2014)
29. Gamo, Y., Nagashima, A., Wakabayashi, M., Terai, M., Oshima, C.: Atomic structure of monolayer graphite formed on Ni(111). Surf. Sci. **374**, 61–64 (1997)
30. Giovannetti, G., Khomyakov, P.A., Brocks, G., Karpan, V.M., van den Brink, J., Kelly, P.J.: Doping graphene with metal contacts. Phys. Rev. Lett. **101**, 026803 (2008)

31. Gui, G., Li, J., Zhong, J.: Band structure engineering of graphene by strain: first-principles calculations. Phys. Rev. B **78**, 075435 (2008)
32. Hwang, E.H., Das Sarma, S.: Screening, Kohn Anomaly, Friedel Oscillation, and RKKY interaction in bilayer graphene. Phys. Rev. Lett. **101**, 156802 (2008)
33. Jhon, Y.I., Zhu, S.-E., Ahn, J.-H., Jhon, M.S.: The mechanical responses of tilted and non-tilted grain boundaries in graphene. Carbon **50**, 3708–3716 (2012)
34. Jiang, J.-W., Wang, J.-S., Li, B.: Young's modulus of graphene: a molecular dynamics study. Phys. Rev. B **80**, 113405 (2009)
35. Kalosakas, G., Lathiotakis, N.N., Galiotis, C., Papagelis, K.: In-plane force fields and elastic properties of graphene. J. Appl. Phys. **113**, 134307 (2013)
36. Kam, K., Scarpa, F., Adhikari, S. & Chowdhury, R.: Graphene nanofilm as pressure and force sensor: a mechanical analysis. Physica Status Solidi (b) **250**, 2085–2089 (2013)
37. Kilian, O., Allan, G., Wirtz, L.: Near Kohn anomalies in the phonon dispersion relations of lead chalcogenides. Phys. Rev. B **80**, 245208 (2009)
38. Kohn, W.: Image of the Fermi surface in the vibration spectrum of a metal. Phys. Rev. Lett. **2**, 393–394 (1959)
39. Kudin, K.N., Scuseria, G.E., Yakobson, B.I.: C2F, BN, and C nanoshell elasticity from ab initio computations. Phys. Rev. B **64**, 235406 (2001)
40. Kwon, S.Y., Ciobanu, C.V., Petrova, V., Shenoy, V.B., Bareño, J., Gambin, V., Petrov, I., Kodambaka, S.: Growth of semiconducting graphene on palladium. Nano Lett. **9**, 3985–3990 (2009)
41. Lazzeri, M., Mauri, F.: Nonadiabatic Kohn anomaly in a doped graphene monolayer. Phys. Rev. Lett. **97**, 266407 (2006)
42. Lazzeri, M., Piscanec, S., Mauri, F., Ferrari, A.C., Robertson, J.: Electron transport and hot phonons in Carbon nanotubes. Phys. Rev. Lett. **95**, 236802 (2005)
43. Lee, C., Wei, X., Kysar, J.W., Hone, J.: Measurement of the elastic properties and intrinsic strength of monolayer graphene. Science **321**, 385–388 (2008)
44. Lee, G.-H., Cooper, R.C., An, S.J., Lee, S., van der Zande, A., Petrone, N., Hammerberg, A. G., Lee, C., Crawford, B., Oliver, W., Kysar, J.W., Hone, J.: High-Strength chemical-vapor–deposited graphene and grain boundaries. Science **340**, 1073–1076 (2013)
45. Lin, Z., Zhigilei, L.V., Celli, V.: Electron-phonon coupling and electron heat capacity of metals under conditions of strong electron-phonon nonequilibrium. Phys. Rev. B **77**, 075133 (2008)
46. Liu, F., Ming, P., Li, J.: Ab initio calculation of ideal strength and phonon instability of graphene under tension. Phys. Rev. B **76**, 064120 (2007)
47. Lu, J.P.: Elastic properties of Carbon nanotubes and nanoropes. Phys. Rev. Lett. **79**, 1297 (1997)
48. Michel, K.H., Verberck, B:. Theory of the elastic constants of graphite and graphene. Physica Status Solidi (b) **245**, 2177–2180 (2008)
49. Milošević, I., Kepčija, N., Dobardžić, E., Damnjanović, M., Mohr, M., Maultzsch, J., Thomsen, C.: Kohn anomaly in graphene. Mater. Sci. Eng B-Adv. Funct. Solid-State Mater. **176**, 510–511 (2011)
50. Milošević, I., Kepčija, N., Dobardžić, E., Mohr, M., Maultzsch, J., Thomsen, C., Damnjanović, M.: Symmetry based analysis of the Kohn anomaly and electron-phonon interaction in graphene and carbon nanotubes. Phys. Rev. B **81**, 233410 (2010)
51. Miniussi, E., Pozzo, M., Baraldi, A., Vesselli, E., Zhan, R.R., Comelli, G., Menteş, T.O., Niño, M.A., Locatelli, A., Lizzit, S., Alfè, D.: Thermal stability of corrugated epitaxial graphene grown on Re(0001). Phys. Rev. Lett. **106**, 216101 (2011)
52. Novoselov, K.S., Geim, A.K., Morozov, S.V., Jiang, D., Katsnelson, M.I., Grigorieva, I.V., Dubonos, S.V., Firsov, A.A.: Two-dimensional gas of massless Dirac fermions in graphene. Nature **438**, 197–200 (2005)
53. Oshima, C., Aizawa, T., Souda, R., Ishizawa, Y., Sumiyoshi, Y.: Surface phonon dispersion curves of graphite (0001) over the entire energy region. Solid State Commun. **65**, 1601–1604 (1988)

54. Pan, Y., Zhang, H., Shi, D., Sun, J., Du, S., Liu, F., Gao, H.-J.: Highly ordered, millimeter-scale, continuous, single-crystalline graphene monolayer formed on Ru (0001). Adv. Mater. **21**, 2777–2780 (2009)
55. Park, S., An, J., Suk, J.W., Ruoff, R.S.: Graphene-based actuators. Small **6**, 210–212 (2010)
56. Peng, Q., Liang, C., Ji, W., De, S.: A theoretical analysis of the effect of the hydrogenation of graphene to graphane on its mechanical properties. Phys. Chem. Chem. Phys. **15**, 2003–2011 (2013)
57. Perebeinos, V., Tersoff, J.: Valence force model for phonons in graphene and carbon nanotubes. Phys. Rev. B **79**, 241409 (2009)
58. Piscanec, S., Lazzeri, M., Mauri, F., Ferrari, A.C., Robertson, J.: Kohn Anomalies and electron-phonon interactions in graphite. Phys. Rev. Lett. **93**, 185503 (2004)
59. Piscanec, S., Lazzeri, M., Robertson, J., Ferrari, A.C., Mauri, F.: Optical phonons in carbon nanotubes: Kohn anomalies, Peierls distortions, and dynamic effects. Phys. Rev. B **75**, 035427 (2007)
60. Pletikosić, I., Kralj, M., Pervan, P., Brako, R., Coraux, J., N'Diaye, A.T., Busse, C., Michely, T.: Dirac cones and minigaps for graphene on Ir(111). Phys. Rev. Lett. **102**, 056808 (2009)
61. Politano, A., Borca, B., Minniti, M., Hinarejos, J.J., Vázquez De Parga, A.L., Farías, D., Miranda, R.: Helium reflectivity and Debye temperature of graphene grown epitaxially on Ru (0001). Phys. Rev. B **84**, 035450 (2011a)
62. Politano, A., Chiarello, G.: Probing Young's modulus and Poisson's ratio in graphene/metal interfaces and graphite: a comparative study. Nano Res. **8**, 1847–1856 (2015)
63. Politano, A., de Juan, F., Chiarello, G., Fertig, H.A.: Emergence of an out-of-plane optical phonon (ZO) Kohn anomaly in quasifreestanding epitaxial graphene. Phys. Rev. Lett. **115**, 075504 (2015)
64. Politano, A., Marino, A.R., Campi, D., Farías, D., Miranda, R., Chiarello, G.: Elastic properties of a macroscopic graphene sample from phonon dispersion measurements. Carbon **50**, 4903–4910 (2012)
65. Politano, A., Marino, A.R., Formoso, V., Chiarello, G.: Evidence of Kohn anomalies in quasi-freestanding graphene on Pt(111). Carbon **50**, 734–736 (2012)
66. Politano, A., Marino, A.R., Formoso, V., Farías, D., Miranda, R., Chiarello, G.: Evidence for acoustic-like plasmons on epitaxial graphene on Pt(111). Phys. Rev. B **84**, 033401 (2011)
67. Reddy, C.D., Ramasubramaniam, A., Shenoy, V.B., Zhang, Y.-W.: Edge elastic properties of defect-free single-layer graphene sheets. Appl. Phys. Lett. **94**, 101904 (2009)
68. Reddy, K.M., Gledhill, A.D., Chen, C.-H., Drexler, J.M., Padture, N.P.: High quality, transferrable graphene grown on single crystal Cu(111) thin films on basal-plane sapphire. Appl. Phys. Lett. **98**, 113117 (2011)
69. Reynaud, C., Sommer, F., Quet, C., el Bounia, N., Duc, T.M.: Quantitative determination of Young's modulus on a biphase polymer system using atomic force microscopy. Surf. Interface Anal. **30**, 185–189 (2000)
70. Ruiz-Vargas, C.S., Zhuang, H.L., Huang, P.Y., van der Zande, A.M., Garg, S., McEuen, P.L., Muller, D.A., Hennig, R.G., Park, J.: Softened elastic response and unzipping in chemical vapor deposition graphene membranes. Nano Lett. **11**, 2259–2263 (2011)
71. Sánchez-Portal, D., Artacho, E., Soler, J.M., Rubio, A., Ordejón, P.: Ab initio structural, elastic, and vibrational properties of carbon nanotubes. Phys. Rev. B **59**, 12678 (1999)
72. Sasaki, K., Yamamoto, M., Murakami, S., Saito, R., Dresselhaus, M.S., Takai, K., Mori, T., Enoki, T., Wakabayashi, K.: Kohn anomalies in graphene nanoribbons. Phys. Rev. B **80**, 155450 (2009)
73. Scarpa, F., Adhikari, S., Srikantha Phani, A.: Effective elastic mechanical properties of single layer graphene sheets. Nanotechnology **20**, 065709 (2009)
74. Scharfenberg, S., Rocklin, D.Z., Chialvo, C., Weaver, R.L., Goldbart, P.M., Mason, N.: Probing the mechanical properties of graphene using a corrugated elastic substrate. Appl. Phys. Lett. **98**, 091908 (2011)

75. Seldin, E.J., Nezbeda, C.W.: Elastic constants and electron-microscope observations of neutron-irradiated compression-annealed pyrolytic and single-crystal graphite. J. Appl. Phys. **41**, 3389–3400 (1970)
76. Shikin, A.M., Farías, D., Adamchuk, V.K., Rieder, K.H.: Surface phonon dispersion of a graphite monolayer adsorbed on Ni(111) and its modification caused by intercalation of Yb, La and Cu layers. Surf. Sci. **424**, 155–167 (1999)
77. Sicot, M., Bouvron, S., Zander, O., Rudiger, U., Dedkov, Y.S., Fonin, M.: Nucleation and growth of nickel nanoclusters on graphene Moiré on Rh(111). Appl. Phys. Lett. **96**, 3–093115 (2010)
78. Sutter, P., Sadowski, J.T., Sutter, E.: Graphene on Pt(111): growth and substrate interaction. Phys. Rev. B **80**, 245411 (2009)
79. Talwar, D.N., Vandevyver, M., Kunc, K., Zigone, M.: Lattice dynamics of zinc chalcogenides under compression: phonon dispersion, mode Grüneisen, and thermal expansion. Phys. Rev. B **24**, 741–753 (1981)
80. Tomasetti, E., Legras, R., Nysten, B.: Quantitative approach towards the measurement of polypropylene/(ethylene-propylene) copolymer blends surface elastic properties by AFM. Nanotechnology **9**, 305 (1998)
81. TSE, W.K., Hu, B.Y.K., Das Sarma, S.: Chirality-induced dynamic Kohn anomalies in graphene. Phys. Rev. Lett. **101**, 066401 (2008)
82. Wagner, P., Ivanovskaya, V.V., Rayson, M.J., Briddon, P.R., Ewels, C.P.: Mechanical properties of nanosheets and nanotubes investigated using a new geometry independent volume definition. J. Phys.: Condens. Matter **25**, 155302 (2013)
83. Wang, B., Günther, S., Wintterlin, J., Bocquet, M.-L.: Periodicity, work function and reactivity of graphene on Ru(0001) from first principles. New J. Phys. **12**, 043041 (2010)
84. Wang, C.G., Lan, L., Liu, Y.P., Tan, H.F.: Multiple component correlation model for elastic modulus of single layer graphene sheets. Physica E **56**, 372–376 (2014)
85. Wang, R., Wang, S., Wu, X., Liang, X.: First-principles calculations on third-order elastic constants and internal relaxation for monolayer graphene. Physica B **405**, 3501–3506 (2010)
86. Wang, Y., Yang, R., Shi, Z., Zhang, L., Shi, D., Wang, E., Zhang, G.: Super-elastic graphene ripples for flexible strain sensors. ACS Nano **5**, 3645–3650 (2011)
87. Wintterlin, J., Bocquet, M.L.: Graphene on metal surfaces. Surf. Sci. **603**, 1841–1852 (2009)
88. XIAO, J.R., Staniszewski, J., Gillespie Jr, J.W.: Fracture and progressive failure of defective graphene sheets and carbon nanotubes. Compos. Struct. **88**, 602–609 (2009)
89. Zakharchenko, K.V., Katsnelson, M.I., Fasolino, A.: Finite temperature lattice properties of graphene beyond the quasiharmonic approximation. Phys. Rev. Lett. **102**, 046808 (2009)
90. Zhang, Y.Y., Wang, C.M., Cheng, Y., Xiang, Y.: Mechanical properties of bilayer graphene sheets coupled by sp3 bonding. Carbon **49**, 4511–4517 (2011)
91. Zhou, G., Duan, W., Gu, B.: First-principles study on morphology and mechanical properties of single-walled carbon nanotube. Chem. Phys. Lett. **333**, 344–349 (2001)
92. Zhou, J., Huang, R.: Internal lattice relaxation of single-layer graphene under in-plane deformation. J. Mech. Phys. Solids **56**, 1609–1623 (2008)
93. Zhou, L., Wang, Y., Cao, G.: Elastic properties of monolayer graphene with different chiralities. J. Phys.: Condens. Matter **25**, 125302 (2013)

Ab Initio Calculations and Kinetic Process Simulations of Nitrogen-Doped Graphene

I. Deretzis and A. La Magna

Abstract The precise control of the doping characteristics of graphene-based systems is a key parameter for successful integration in devices and applications. Doping can tune graphene's carrier density, which is important for the optimization of graphene-based flexible and transparent electrodes. Moreover, if doping is confined within one out of the two equivalent graphene sublattices, considerable band gaps can open that could serve for switching devices on and off. We use the density functional theory to explore the conditions of sublattice symmetry breaking in nitrogen-doped graphene. We show that the nucleation of graphene grains and their growth through the propagation of unpassivated zigzag edges could give rise to a highly unbalanced doping, due to exothermic substitutions on edge sites. Based on our *ab initio* results, we calibrate an event-driven kinetic Monte Carlo model that simulates the growth kinetics and investigates the N incorporation mechanism in real-process conditions. We demonstrate that the degree of symmetry breaking strongly depends on the growing conditions.

1 Introduction

The controlled growth and the precise manipulation of graphene systems constitute the two important requirements for a future industrial use of graphene. The first one is fundamental for the reproducibility of graphene devices, whereas the second can engineer the necessary functionality/versatility for a wide range of applications. To date, a key process for the controlled production of graphene is the chemical vapor deposition (CVD) of organic precursors on metal substrates [1, 2]. Such process can be further combined with the in situ doping of graphene by atoms of different species, which could add further functionality on the graphene films. A particular

I. Deretzis (✉) · A. La Magna
Istituto per la Microelettronica E Microsistemi (CNR-IMM),
Z.I. VIII strada 5, 95121 Catania, Italy
e-mail: ioannis.deretzis@imm.cnr.it

A. La Magna
e-mail: antonino.lamagna@imm.cnr.it

V. Morandi and L. Ottaviano (eds.), *GraphITA*, Carbon Nanostructures,
DOI 10.1007/978-3-319-58134-7_5

case of such process is the functionalization of graphene with either nitrogen [3–5] or boron [6, 7]. These two elements have the capability of acquiring an sp^2-type hybridization like carbon. Hence, when they substitute the C atoms in the graphene film they maintain a planar geometry, having a minimal impact on the intrinsic electrical properties of the material [8]. At the same time, they give rise to a tunable doping as well as to energy-specific electronic alterations. These characteristics could find implementation in devices of various types; here we point out touch screens and transparent electrodes based on graphene (where a reduction of the sheet resistance is necessary) and transistors (where mobility or intrinsic band gaps are required in order to switch-off the device).

The synthesis of nitrogen-doped graphene has also revealed some peculiar topological features regarding the presence of substitutional nitrogen atoms in one out of the two graphene sublattices [4, 5]. The peculiarity stands at the fact that the two constituent sublattices that form the honeycomb structure are perfectly equivalent. This aspect is indeed of particular importance, as sublattice symmetry-breaking could ideally give rise to a real band-gap in the low-energy electronic spectrum of graphene, making it suitable for switching applications. Curiously, the phenomenon has been reported only for CVD processes, where doping takes place simultaneously along with the growth of graphene, and not for *ex situ* manipulations (e.g., when the two-dimensional graphene sheet is post-processed with nitrogen plasma).

This paper intends to explore from a theoretical perspective the origins of the sublattice asymmetry in the doping of graphene with nitrogen as well as the thermodynamic conditions of growth that could allow the manifestation of such characteristic. To this end, we employ a combined theoretical/computational approach within the density functional theory (DFT) and the kinetic lattice Monte Carlo (KLMC) schemes. Our results show that substitutional nitrogen atoms can occupy positions that belong to a single sublattice when they are incorporated on the zigzag edges of graphene grains, either directly or through spontaneous exothermic substitutions of carbon atoms. We note that the grain-growth model is characteristic of graphene CVD-based processes [9]. We moreover argue that in order to take place, such phenomenon requires particular thermodynamic conditions of growth that favor the defect-free propagation of zigzag edges for the growing graphene grains. We show that such a thermodynamic equilibrium is not easily achievable, as the required temperatures may give rise to diffusion effects for graphene vacancies, which, in turn, may compromise the sublattice asymmetry feature. We finally discuss why process simulations are fundamental for the understanding of the kinetic processes that lead to graphene formation.

The paper is organized as follows: In Sect. 2 we give an overview of the computational methodology. Section 3 discusses the DFT results regarding doping in two-dimensional (Sect. 3.1) and zero-dimensional (Sect. 3.2) graphene, while Sect. 4 shows kinetic results based on the kinetic Monte Carlo scheme. Finally, in Sect. 5 we briefly discuss our results.

2 Methodology

The study focuses on the energetics and kinetics of nitrogen-doped graphene within a multiscale computational methodology. Structural and electronic properties are studied within the DFT. We use the SIESTA code [10], localized atomic orbitals for the expansion of the Kohn-Sham wave-functions and norm-conserving pseudopotentials for the description of the core electrons. The local density approximation [11] is applied for the description of the exchange-correlation functional. We consider two types of graphene systems; periodic graphene supercells (i.e., two-dimensional graphene) and hexagonal graphene quantum dots (i.e., zero-dimensional graphene). In the first case, we construct 18×18 supercells with a total number of 648 atoms. Due to the increased computational load, we perform Γ-point calculations and use a double-ζ basis set for both C and N. In the case of the hexagonal quantum dots we consider clusters of 96 up to 216 atoms and a basis set that also includes polarization orbitals. In both calculations atoms are allowed to relax until forces are less than 0.04 eV/Å.

The DFT results are used to partially calibrate a KLMC algorithm [12] that simulates the CVD growth of graphene, considering a synchronous functionalization with nitrogen (i.e., the algorithm considers a CVD process where the precursors contain both C and N). Moreover, the KLMC model is capable of simulating the thermal kinetics of vacancy-type defects, according to the parameterization presented in Ref. [13]. The evolution of the system takes place trough a configuration-dependent activation energy model, which is based on a sequence of stochastic events occurring with a rate ν_i. The event rate ν_i is correlated to the occurrence probability of a particular event and is calculated according to transition state theory by a Boltzmann weight

$$\nu_i = \frac{1}{\tau_i} = \nu_0 exp[-E_i/k_B T]. \tag{1}$$

Here k_B is the Boltzmann constant, T the absolute temperature, E_i is the activation energy and ν_0 a prefactor that includes the contributions that are not included in the energy barrier. This rate form implies that events that are more likely to occur will happen more frequently. The energy barrier E_i for the migration of a particular atom has been set to depend on the initial local configuration. Such interaction energy has been further modeled to depend on the number of occupied sites at nearest-neighbor and next-nearest-neighbor sites. Thus, rates depend only on the coordination of the C or N atoms. We moreover note that the calibration of the KLMC model accurately reproduces the energy gain obtained when two single-vacancies merge to form a divacancy, or when a nitrogen atom merges with a single vacancy in the extremely stable pyridinic configuration.

The advantage of our KLMC algorithm stands at the possibility to simulate long-time processes (comparable to those of the experimental case), while preserving the atomic character of the simulation. We note that this aspect is unique within the kinetic computational framework, as classical molecular dynamics simulations have a temporal limit that does not exceed the μs scale, when usually growth processes range from minutes to hours.

3 DFT Results

3.1 Nitrogen Doping in Two-Dimensional Graphene

In this paragraph, we discuss the aspect of sublattice symmetry breaking in N-doped graphene, starting from energetic considerations. Figure 1 shows a graphene honeycomb lattice, indicating the atoms that belong to the two equivalent trigonal sublattices. A first check for the determination of the preference of substitutional N atoms to occupy positions of one out of the two sublattices is to understand if there are energetic preferences for such behavior. For this purpose we define the interaction energy for two substitutional N atoms as a function of their distance, as

$$E_{int}(d) = E_{2N}(d) - (2 \times E_N - E_{tot}), \tag{2}$$

where E_{2N} is the total energy of a supercell containing 2 substitutional N atoms at distance d, E_N is the energy of the same supercell containing a single N atom and E_{tot} is the energy of the supercell without N additions. Equation 2 requires that the size of the used supercell is big enough in order to minimize the interaction between the periodic replicas of the defects in neighboring cells. This aspect is particularly important in the case of substitutional N, as the latter acts as a donor that gives rise to a delocalized charge around the N site with an exponential tail that decays at ~2.2 nm from the defect origin [14]. Here we consider supercells with a 18 × 18 periodic-

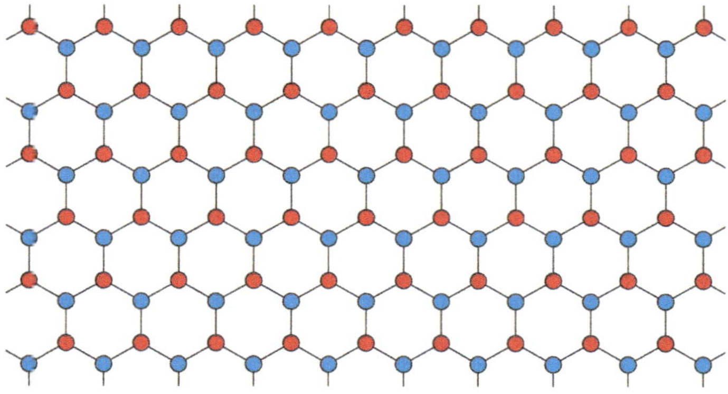

Fig. 1 Schematic view of the honeycomb graphene structure, highlighting the presence of the two equivalent trigonal sublattices

Fig. 2 Interaction energy E_{int} for two substitutional nitrogen atoms in graphene

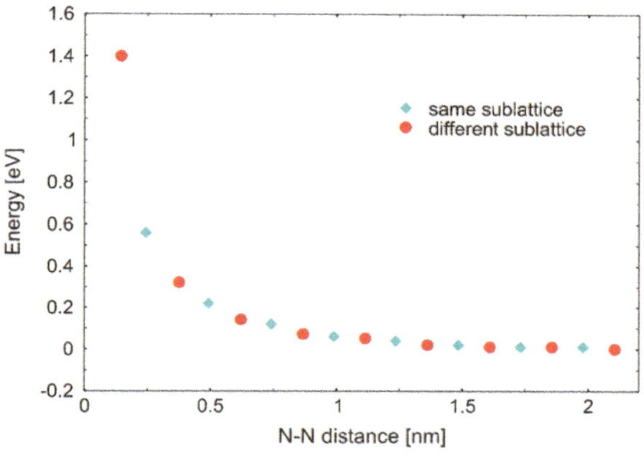

ity, which reduce the error in the energy calculation below 0.1 eV. The results of E_{2N} as a function of distance of the two substitutional atoms can be seen in Fig. 2. We note that the interaction energy has typical repulsive characteristics, which can be attributed to Coulomb interactions between the two electron concentrations around the substitutional nitrogens. We moreover notice an absence of any energetic preference when the two N atoms stand on the same or at different sublattices. Hence, the energetic picture obtained from the DFT calculations shows that substitutional N defects in two-dimensional graphene have no sublattice preference, while they increase the energy of the system when they appear within their range of interaction (~4.4 nm). However, these results are in clear contradiction with the experimental studies of CVD growth that have evidenced both a sublattice preference [4] and an often presence of N atoms in next-nearest-neighbor configurations [5]. This experimental-theoretical discrepancy indicates that the two-dimensional model for the N-doping of graphene is not suitable for the description of CVD-based processes. This aspect will be discussed in the next paragraph.

3.2 *Nitrogen Doping in Zero-Dimensional Graphene Grains*

As stated earlier, CVD-based processes of graphene synthesis are driven by the nucleation and growth of graphene grains that slowly expand over the metallic substrate. Such grains often obtain a hexagonal shape and their edges typically have zigzag-type terminations [9]. For our calculations we consider the 96-atom hexagonal graphene grain seen in Fig. 3a, noting that similar results have been obtained for bigger grains (with a total number of 216 atoms). We first check the total energy of such zero-dimensional system in the presence of a single substitutional N defect at different positions, considering the center-to-edge path seen in Fig. 3a. The results (Fig. 3b) show that the total energy of the system significantly lowers when the N atoms occupy edge positions. In particular we observe a ~2 eV energy gain when

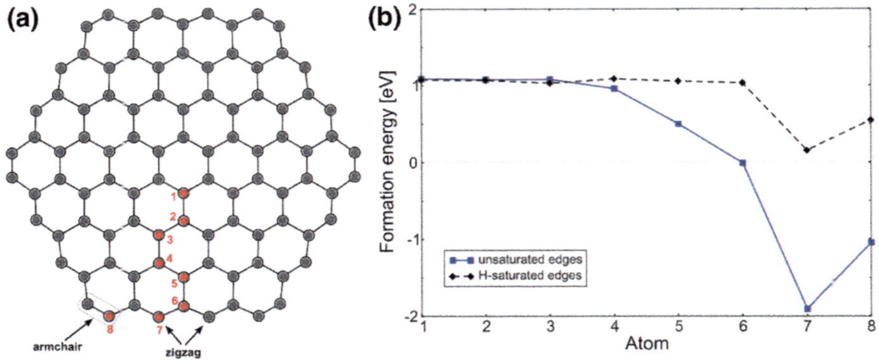

Fig. 3 **a** Schematic view of a 96-atom hexagonal graphene grain used in the calculations, indicating armchair and zigzag-type edges. **b** Formation energy for a substitutional N atom at different positions of the graphene grain

the N atom occupies an armchair-type edge and a ~3 eV energy gain when it occupies a zigzag-type edge. This aspect denotes the strong preference of N atoms for the pyridinic configuration, where the missing C-N bond is replaced by a pair of N valence electrons which fully recovers the sp^2-type hybridization. We calculate the formation energy of N substitutional defects within such hexagonal grains as

$$E_f = E_N - E_{bulk} - \mu_C + \mu_N, \quad (3)$$

where E_N is the energy of the system with the N substitutional atom, E_{bulk} is the energy of the non-defected graphene grain, μ_C is the chemical potential of carbon (with graphite as a reference) and μ_N the chemical potential of N (considering N_2 as a reference). Results (Fig. 3b) show that the substitution of C atoms with N should be an exothermic process (i.e., the substitution of C by N should be spontaneous on edge sites). We also point out that this characteristic is valid only for bare edges, as the replica of the previous calculations considering H-passivated edges strictly gives rise to positive formation energies (Fig. 3b). Hence, the zero-dimensional model results in a new interpretation for the sublattice symmetry-breaking effect seen in CVD processes, considering the gradual expansion of graphene grains with zigzag edges: Since all zigzag sites of a single edge belong to the same sublattice and hexagonal grains are composed of six triangles with edges of alternating sublattice, this scheme leads to triangular domains (within the same grain) with a sublattice preference for N atoms, delimited by straight lines. Such model is in good agreement with the experiments [4] However, its applicability should ideally be validated by considering the influence of the process parameters and growth conditions on the incorporation of the N atoms on the grain edges. This can be achieved through an advanced KLMC implementation of the graphene growth process, discussed below.

4 Process Simulation for Nitrogen-Doped Graphene

The picture extracted from the total energy calculations indicates a clear sublattice preference for nitrogen atoms during a growth process that evolves through the propagation of ideal zigzag-type edges. The question that arises is if such characteristic can be maintained in real-process conditions, where thermodynamic (temperature, pressure) and morphological (terraces and steps of the Cu substrate) parameters play a crucial role in the formation and expansion of the graphene grains. To this end, process simulations that account for these issues are necessary for a more realistic evaluation of the sublattice symmetry problem. Here, we attempt a first-order consideration of such aspect through kinetic lattice Monte Carlo simulations, considering ideal substrates, while varying the process temperature and the partial pressures of the precursors. The latter are effectively considered by tuning the nitrogen fluxes in the simulation settings.

Figure 4a shows a KLMC simulation of N-doped graphene growth after the initial stages of nucleation, for $T = 800\,°\text{C}$. A first observation is that the simulation reproduces the hexagonal shape of the actual CVD-based process for graphene grains. However, a closer look on the edges shows that notwithstanding the prevalent zigzag character of edge terminations, there exists a clear edge roughness with armchair- and Klein-type inclusions. Such roughness is responsible for the limitation of the sublattice preference for N dopants, as seen in Fig. 4a (where substitutional atoms in the two sublattices are colored with red and blue, respectively). Hence, a strategy to overcome this issue could be to better control the reconstruction of grain edges in order to maintain in the better way possible the zigzag character during growth. Our simulations show that such situation is possible with the increase of the process temperature. However, when calculating the sublattice asymmetry degree for

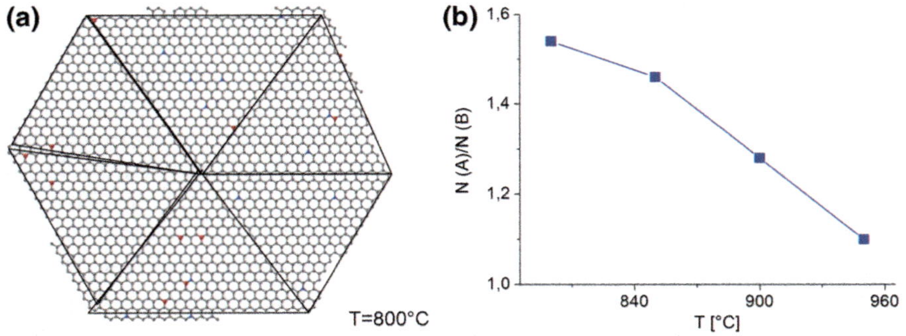

Fig. 4 **a** Snapshot of an N-doped nanograin of graphene after the early stages of a KLMC simulated growing process on a Cu substrate. The effective atomic flux is $\Phi = 10^{18}$ at/cm^{2}s, while the ratio between the flux of N atoms and C atoms is equal to 0.2. The approximated triangle-like domains are shown. *Blue* and *red* colors indicate atoms in different sublattices. **b** Sublattice asymmetry degree as a function of the temperature calculated in large ($\sim\mu$m) domains, where the initial configuration is a hexagonal grain with ideal zigzag borders

N substitutional atoms as a function of the temperature (see Fig. 4b) we observe the opposite picture, i.e., a decrease of the sublattice asymmetry, notwithstanding the fact that edge roughness indeed lowers. The origin of such discrepancy lies at the kinetics of point defects (i.e., vacancies) in the graphene lattice, which our KLMC algorithm fully considers. Vacancies are generated at the growing edges due to the imperfect reconstruction of the dangling bonds when N atoms are incorporated to the grain. By increasing the temperature above $T = 800\,^{\circ}\mathrm{C}$ vacancy diffusion and merging phenomena tend to increase, which in turn, indirectly influences the position of the N dopants and, hence, their presence in one out of the two graphene sublattices. Such paradigm is characteristic for the importance of the kinetic aspects of growth, in addition to the energetic aspect, for the determination of the topological characteristics of graphene grown through CVD processes. In this sense, kinetic process simulations could be fundamental not only for assisting the experimental procedure, but also for understanding the physical origins of the obtained geometries. We finally note that a further improvement of the presented model could include the active presence of the Cu substrate (considering also the kinetics of surface Cu atoms, see Ref. [15] for a similar formulation in the case of graphene on SiC) and a better calibration of the process parameters (i.e., activation energies, frequency prefactors) in order to describe more accurately the laboratory-specific characteristics of the experimental setup.

5 Discussion

The peculiarity of the sublattice symmetry-breaking effect during the nitrogen doping of graphene in CVD-based process as well as the possibilities that this opens for the manufacturing of novel graphene devices [14] render the understanding of the phenomenon particularly interesting. In this work we have employed density functional theory calculations and kinetic lattice Monte Carlo simulations in order to investigate this issue, considering the energetic, kinetic and thermodynamic aspects of N-doping during CVD growth. Our DFT calculations have indicated that this effect is correlated with the incorporation of N atoms during the thermal expansion of graphene grains, which is typical of the CVD growth mechanism. In particular we found that the reactivity of propagating zigzag edges with nitrogen is the principal motivation for this sublattice symmetry-breaking mechanism. Such purely-energetic model has been further validated by appropriately calibrated KLMC simulations that also account for kinetic and thermodynamic aspects of the growth process. The KLMC results showed that the degree of sublattice symmetry-breaking strongly depends on the process conditions. In particular, we have evidenced that the manifestation of the phenomenon requires the propagation of almost ideal, defect-free zigzag edges, which necessitate increased process temperatures. The latter can additionally trigger vacancy diffusion phenomena and related vacancy-N interaction mechanisms that, in turn, should compromise the sublattice asymmetry. The combined results indicate that process simulations are fundamental for the understanding of graphene-based processes, the quality of the resulting material and consequently its structural and electronic characteristics.

Acknowledgements The authors would like to acknowledge the FLAG-ERA GraNitE project for partial financial support.

References

1. Reina, A., Jia, X., Ho, J., Nezich, D., Son, H., Bulovic, V., Dresselhaus, M.S., Kong, J.: Large area, few-layer graphene films on arbitrary substrates by chemical vapor deposition. Nano Lett. **9**, 30–35 (2008)
2. Li, X., Magnuson, C.W., Venugopal, A., Tromp, R.M., Hannon, J.B., Vogel, E.M., Colombo, L., Ruoff, R.S.: Large-area graphene single crystals grown by low-pressure chemical vapor deposition of methane on copper. J. Am. Chem. Soc. **133**, 2816–2819 (2011)
3. Usachov, D., Vilkov, O., Gruneis, A., Haberer, D., Fedorov, A., Adamchuk, V.K., Preobrajenski, A.B., Dudin, P., Barinov, A., Oehzelt, M., Laubschat, C.: Nitrogen-doped graphene: efficient growth, structure, and electronic properties. Nano Lett. **11**, 5401–5407 (2011)
4. Zabet-Khosousi, A., Zhao, L., Plov, L., Hybertsen, M.S., Reichman, D.R., Pasupathy, A.N., Flynn, G.W.: Segregation of sublattice domains in nitrogen-doped graphene. J. Am. Chem. Soc. **136**, 1391–1397 (2014)
5. Lv, R., Li, Q., Botello-Mndez, A.R., Hayashi, T., Wang, B., Berkdemir, A., Hao, Q., Elas, A.L., Cruz-Silva, R., Gutirrez, H.R., Kim, Y.A.: Nitrogen-doped graphene: beyond single substitution and enhanced molecular sensing. Scientific Rep. **2**, 586 (2012)
6. Zhao, L., Levendorf, M., Goncher, S., Schiros, T., Palova, L., Zabet-Khosousi, A., Rim, K.T., Gutierrez, C., Nordlund, D., Jaye, C., Hybertsen, M.: Local atomic and electronic structure of boron chemical doping in monolayer graphene. Nano Lett. **13**, 4659–4665 (2013)
7. Gebhardt, J., Koch, R.J., Zhao, W., Hfert, O., Gotterbarm, K., Mammadov, S., Papp, C., Grling, A., Steinrck, H.P., Seyller, T.: Growth and electronic structure of boron-doped graphene. Phys. Review B **87**, 155437 (2013)
8. Lherbier, A., Blase, X., Niquet, Y.M., Triozon, F., Roche, S.: Charge transport in chemically doped 2D graphene. Phys. Review Lett. **101**, 036808 (2008)
9. Yu, Q., Jauregui, L.A., Wu, W., Colby, R., Tian, J., Su, Z., Cao, H., Liu, Z., Pandey, D., Wei, D., Chung, T.F., et al.: Control and characterization of individual grains and grain boundaries in graphene grown by chemical vapour deposition. Nat. Mat. **10**, 443–449 (2011)
10. Soler, J.M., Artacho, E., Gale, J.D., Garca, A., Junquera, J., Ordejn, P., Snchez-Portal, D.: The SIESTA method for ab initio order-N materials simulation. J. Phys. Condens. Matter **14**, 2745 (2002)
11. Perdew, J.P., Zunger, A.: Self-interaction correction to density-functional approximations for many-electron systems. Phys. Review B **23**, 5048 (1981)
12. La Magna, A., Coffa, S., Colombo, L.: Role of extended vacancy-vacancy interaction on the ripening of voids in silicon. Phys. Review Lett. **82**, 1720 (1999)
13. Parisi, L., Di Giugno, R., Deretzis, I., Angilella, G.G.N., La Magna, A.: Kinetic Monte Carlo simulations of vacancy evolution in graphene. Mat. Sci. Semicond. Process. **42**, 179–182 (2016)
14. Deretzis, I., La Magna, A.: Origin and impact of sublattice symmetry breaking in nitrogen-doped graphene. Phys. Review B **89**, 115408 (2014)
15. Deretzis, I., La Magna, A.: Simulating structural transitions with Kinetic Monte Carlo: The case of epitaxial graphene on SiC. Phys. Review E **93**, 033304 (2016)

From Point to Line Defects in Two-Dimensional Transition Metal Dichalcogenides: Insights from Transmission Electron Microscopy and First-Principles Calculations

H.-P. Komsa and A.V. Krasheninnikov

Abstract Two-dimensional (2D) transition metal dichalcogenides (TMDs) have recently received great deal of attention due to their unique properties associated with the reduced dimensionality of the system. The properties of these materials have been shown to be affected by atomic defects in the atomic network. The very structure of these materials which are composed from three atomic layers only, combined with dramatic improvements in microscopy techniques, made it possible to study the behavior of defects in these systems with unprecedented accuracy. Various point and line defects were identified, and their effects on the properties of the systems were accessed. It was demonstrated that point defects induced by electron beam irradiation coalesce in line defects, but their quasi-one dimensional atomic structure varies from member to member in the transition metal dichalcogenides family. In this review, we summarize recent experimental and theoretical findings in this area, discuss how the line structures appear due to the agglomeration of point defects, and dwell upon how line defects can be used to engineer properties of 2D TMDs. Finally, we address the challenges in this field and issues which still lack the explanation.

1 Introduction

Following the isolation [1] and synthesis [2] of single sheets of graphene, two-dimensional (2D) transition metal dichalcogenides (TMDs), which sometimes are referred to as inorganic graphenes, with a common structural formula MeX_2, where Me stands for transition metals (Mo, W, Tc, Pd, Pt, etc.), X for chalcogens S, Se, Te, became the subject of intensive research, see Refs. [3–5] for an overview. Contrary

H.-P. Komsa (✉) · A.V. Krasheninnikov
COMP, Department of Applied Physics, Aalto University, P.O. Box 11100, 00076 Aalto, Finland
e-mail: hannu-pekka.komsa@aalto.fi

A.V. Krasheninnikov
Helmholtz-Zentrum Dresden-Rossendorf, Institute of Ion Beam Physics and Materials Research, 01328 Dresden, Germany
e-mail: a.krasheninnikov@hzdr.de

V. Morandi and L. Ottaviano (eds.), *GraphITA*, Carbon Nanostructures,
DOI 10.1007/978-3-319-58134-7_6

to graphene, which is a semimetal, and where the gap in the electronic spectrum can be opened only with considerable difficulties, these materials have diverse electronic properties varying from metals to wide-gap semiconductors. The 2D TMD materials have already shown a good potential for nanoelectronics, photonics, catalysis, and energy applications [3, 5–7] due to a unique combination of electronic, optical, and mechanical properties. Putting aside applications of 2D materials, they proved to be fascinating structures from the viewpoint of fundamental physics of the systems with reduced dimensionality.

In retrospect, it is apparent that the quick progress in the investigation of graphene, hexagonal boron-nitride (h-BN), TMDs, and other 2D materials can be credited in part to the recent immense progress [8–10] in aberration-corrected high-resolution transmission electron microscopy (HR-TEM) and the very nature of any two-dimensional system. Oversimplifying, TEM studies of bulk materials assume fabrication of a thin (yet preserving the morphology of the bulk system) sample, followed by the reconstruction of its atomic structure from the TEM image, which can be referred to as a "projection" of atoms forming the structure to a plane. Except for the exfoliation of the 2D material from its bulk counterpart and its transfer to the TEM grid, investigations of systems composed from a single (or few) atomic layers do not require these steps. Moreover, by focusing electron beam on specific areas (essentially on single atom), and using electron energy loss spectroscopy (EELS), it is possible now to identify impurity atoms in substitutional positions [11] and get the information on the local electronic structure/bonding configurations [10, 12], magnetic states [13], and even phonons [14].

As for the fundamental aspects of materials science and solid-state physics, TEM studies of 2D materials, especially when combined with atomistic simulations, have provided plethora of unique data on the behavior of point and line defects. For example, the migration of vacancies, their coalescence, and transformations to agglomerations of non-hexagonal rings have been observed in graphene in real time with atomic resolution [15–17]. Likewise, the atomic structures of point intrinsic [18–20] and extrinsic [21] defects in 2D TMDs and other inorganic 2D materials such as h-BN [22] or silica bilayers [23] have been obtained, along with deep insights into their behavior

TEM experiments also made it possible to identify the atomic structure and properties of line defects in 2D systems, such as dislocations and grain boundaries, see Ref. [24] for an overview. For example, the birth, motion, interaction, and annihilation of dislocations in graphene has been observed [15] during in-situ experiments under electron beam irradiation.

As for grain boundaries, they are particularly important for understanding the mechanical properties of polycrystalline 2D materials grown by chemical vapor deposition and related techniques, as well as for electronic transport in these systems and even optical characteristics. Grain boundaries normally appear when islands of the growing material with different orientations of the crystal lattice coalesce. Depending on the misorientation angle and growth conditions, the morphology of the boundaries can be quite different. For instance, the grain boundaries in graphene consisting mostly from non-hexagonal rings, including pentagons, heptagons and octagons, can be straight or serpentine [25], depending on the misorientation angle between the grains. The structure of grain boundaries [19, 26–28] in TMDs can also vary: while tilt and mirror-twin boundaries in MoS_2 were reported to consist of 8- and 4-membered rings [28], another study pointed to the existence of 5- and 7-membered rings [27]. Besides, mirror-twin boundaries have been reported to appear in $MoSe_2$ and WSe_2 as a corollary of chalcogen atom deficiency during the growth [29]. In semiconducting TMD, grain boundaries and other one-dimensional defects can be conducting channels [28–34], while electronic transport in the direction perpendicular to the boundaries is normally suppressed. First-principles calculations indicate that some line defects in TMDs can exhibit magnetism [29, 32, 35], thus adding new functionalities to the system.

Moreover, a number of in-situ TEM experiments [29, 36–38] showed that agglomeration of vacancies in TMDs gives rise to formation of line defects and development of grains embedded into a larger grain with a different orientation of the crystal lattice of the material. An interesting and somewhat unexpected observation is that depending on the chemical content of TMDs, line defects with different morphologies are formed. For example, agglomeration of vacancies in MoS_2 gives rise to straight vacancy lines without any substantial local transformations of the atomic structure [36, 39], while the sputtering of chalcogen atoms in WSe_2, WS_2 results in the development of rotational defects and eventually mirror grains with the boundaries consisting of 5- and 8-member rings [37]. At the same time, under similar conditions formation of different line structures in $MoSe_2$ was reported [29, 38].

The goal of this review is to give a brief overview of how line defects appear in 2D TMDs under electron beam, analyze their atomic structures and physical properties, compare formation mechanisms, and outline what is not yet fully understood. Finally, the perspectives for using electron beam for engineering the microstructure of 2D TMDs will be discussed.

2 Development of Vacancy Lines in MoS_2 Sheets

In Ref. [36], mechanically exfoliated MoS_2 flakes were deposited on TEM grids and subjected to electron beam irradiation inside a TEM. The electron beam energy was 80 keV, slightly below the calculated static threshold for knock-on damage in MoS_2 (of 90 keV), with thermal movement of the ions yielding nonzero cross section [18, 40]. It should also be pointed out that other mechanisms of damage production such as ionization, electronic excitations, and beam-mediated chemical etching may have also contributed to appearance of vacancies.

Figure 1a shows the initial TEM image of the MoS_2 sample. It contains a number of vacancies, of which a substantial part is likely created during the transfer of the flake to the TEM grid. During imaging, new vacancies were continuously produced. The atomic structures of single vacancy (one S atom missing) is shown in the inset of panel (a). The diffusion of vacancies was observed, although at a rather slow rate.

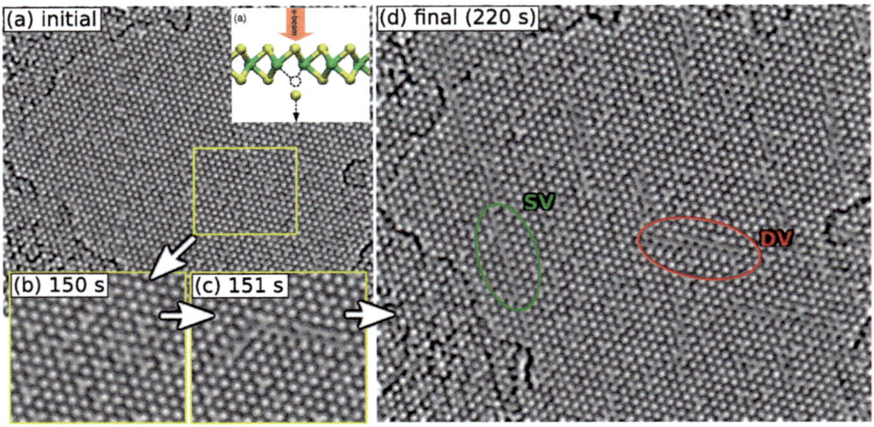

Fig. 1 HR-TEM images showing the formation of line defects under electron irradiation at 80 kV. **a** The initial configuration with a certain number of S vacancies already present. Formation of a sulfur vacancy is schematically shown. **b, c** Intermediate configurations illustrating formation of defect lines through agglomeration of nearby vacancies. **d** The final configuration after 220 s, where a large number of defect lines can be seen. Examples of single vacancy (SV) and double vacancy (DV) lines are highlighted. The shape of the DV line has changed from that shown in panel (**c**). The HR-TEM frames are lowpass filtered in order to reduce Poisson noise. From Ref. [36]

With increasing vacancy concentration, defect agglomeration into vacancy clusters (holes) might be expected, as in many bulk solids, provided that vacancies are mobile enough. However, formation of extended line defects was observed instead. The atomic structure of these lines appears to be adequately described by a row of single vacancies (cf. SV in Fig. 1d), without significant atomic reconstruction. However, the vacancies were also often found to migrate out of the line structure, suggesting that the related energy gain is small.

After high-dose exposure to the electron beam, and further increase in vacancy concentration, the nature of the lines changed. The lines became wider, where, in essence, the lines extended to span two neighboring rows of the S sublattice and, furthermore, the atomic structure underwent significant changes as compared to the single vacancy line (cf. DV in Fig. 1d). Figure 1b, c shows two sequential images of the formation of this kind of vacancy lines. During the formation, the number of single vacancies surrounding the line has dropped significantly, suggesting that the line is formed through agglomeration of these vacancies rather than through production of a large number of new vacancies. The positions of the vacancies far from the line defect remained the same.

The experimental images of the two types of vacancy lines, which can be referred to as single vacancy (SV) and double vacancy (DV) lines, are shown in Fig. 2a, d. In order to determine the underlying atomic structure, first-principles calculations for various atomic models corresponding to vacancy lines were carried out. The opti-

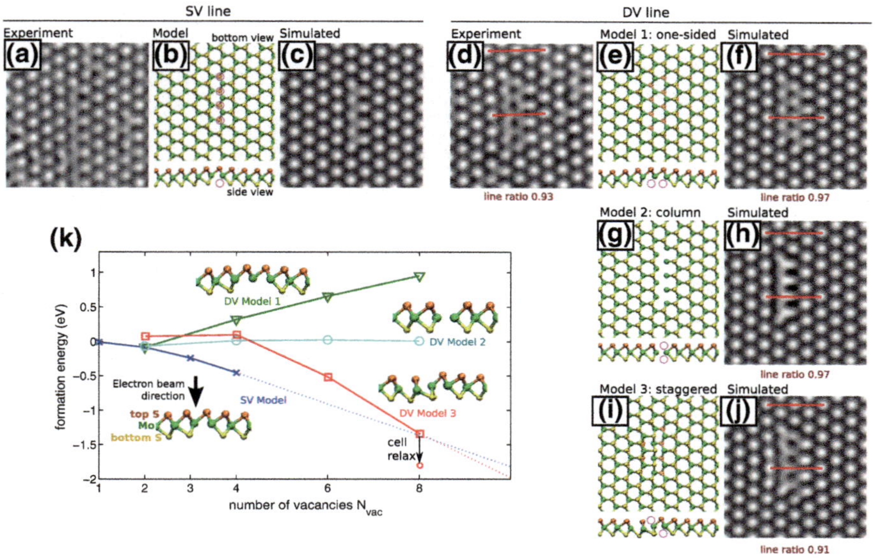

Fig. 2 Comparison of experimental and simulated TEM images. **a, d** Examples of experimental HR-TEM images of the SV and DV lines with filtering applied as in Fig. 1. Various models of the vacancy lines: **b, c** single vacancy line, **e, f** two neighboring vacancy lines in the same S layer, **g, h** two vacancy lines coinciding in top and bottom layers. **i, j** Two vacancy lines in staggered configuration, the atomic structures (**b, e, g, i**) are obtained from DFT calculations and used to simulate TEM images (**c, f, h, j**). Note that the bottom and side views are shown. In order to differentiate between atoms in the two S layers, they are colored differently: *orange* for the top layer and *yellow* for the bottom layer. *Magenta circles* denote the initial vacancy positions. For DV lines, the local shrinkage of the lattice is quantified by a ratio of the length of line segments (*solid red lines*) in pristine lattice and around the defect. **k** Formation energies for all considered models as a function of the number of constituting single vacancies (the number of missing S atoms). The DV lines always have even number of vacancies. The extrapolation to longer lines is also shown. The insets are the side views of the atomic models presented in panels (**b, e, g, i**). From Ref. [36]

mized geometries were used as an input for TEM image simulation (Fig. 2c, f, h, j), which were then compared to the experimental images.

The relative formation energies E_f for these structures are given in Fig. 2k. This quantity was defined as the energy released upon vacancy agglomeration into a line defect with respect to the same number N_{vac} of isolated vacancies:

$$E_f(N_{\text{vac}}) = E_{\text{tot}}(N_{\text{vac}}) + (N_{\text{vac}} - 1)E_{\text{tot}}(\text{pristine}) - N_{\text{vac}}E_{\text{tot}}(\text{SV}) \quad (1)$$

where $E_{\text{tot}}(N_{\text{vac}})$ and $E_{\text{tot}}(\text{SV})$ are the total energy of supercells containing N_{vac} vacancies and a single vacancy, respectively. Negative values of E_f point toward the tendency of vacancies to agglomerate, while positive values indicate that such configuration is energetically unfavorable.

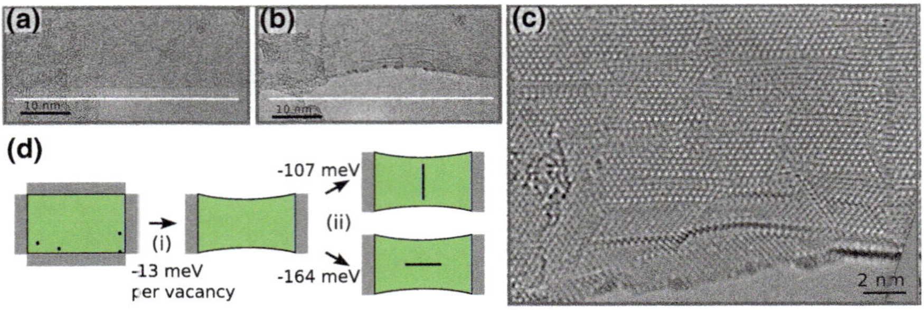

Fig. 3 **a** HR-TEM images of the MoS_2 flake edge. **a** The initial configuration and **b** after 818 s of irradiation. The *white horizontal lines* are guides to the eye. **c** Magnification of the final configuration showing large concentration of lines, predominantly oriented parallel to the edge. **d** A sketch illustrating the formation and energetics of vacancy lines near the MoS_2 flake edge with referential orientation. The energies are normalized to the number of isolated vacancies, and calculated at the 3.1% vacancy concentration. From Ref. [36]

For SV lines, the formation energies show that agglomeration to a line leads to energy gain of about 0.05–0.2 eV per S vacancy, agreeing well with the tendency to line up, and the simulated and experimental images agreed well. As for DV lines the simulations showed that among several possible configurations shown in Fig. 2, only the "staggered" configuration is energetically favorable and also provided the best agreement with the experiment, so that the DV line in experiments were assigned to this structure (model 3).

It was further shown that vacancy lines do not appear until vacancy concentration in the area is sufficiently high, so that there are vacancies at neighboring S sites, and the diffusion barrier of single vacancies then drops from 2.3 down to 0.8 eV, and line defects form.

The formation energy of line defects was theoretically found to strongly depend on mechanical strain. This indicates that external control of strain in the sample can favor formation of lines of particular orientation. Indeed, HR-TEM experiments confirmed that. Figure 3a shows an HR-TEM image taken in the vicinity of the flake edge. On other sides, MoS_2 is clamped to the TEM grid. Initially, the edge is straight. After prolonged electron irradiation, the edge is bent inwards. Most interesting, the majority of the line defects are seen to be directed parallel to the edge. This is due to the possibility to release strain from the direction of the edge, but not from the perpendicular direction.

It was suggested that the full process leading to preferentially aligned line defects is the following, as illustrated in Fig. 3d: (i) With increasing concentration of single vacancies, the lattice may release the stress from the direction of flake edges leaving the sample effectively under uniaxial tensile strain, (ii) Such non-uniform strain favors formation of lines oriented parallel to edges by more than 50 meV per S vacancy, (iii) Formation of DV lines leads to additional contraction of the lattice perpendicular to the lines, but relieves the stress parallel to the line. Thus, active control of strain during electron irradiation should lead to formation of parallel lines.

3 Mirror-Twin Boundaries in $MoSe_2$

Interestingly enough, different structures were observed in the atomic network of $MoSe_2$, a material closely related to MoS_2 in the TMD "family". These are inversion domains separated by mirror twin boundaries (MTBs), as shown in Fig. 4. Due to the symmetry of the underlying lattice, the areas with the mirror symmetry also correspond to a 60° rotation of the lattice.

MTBs with different atomic structures with Mo and Se edges are possible, with the one shown in Fig. 4j having the lowest formation energy among all possible twin boundary structures, as first-principles simulations indicate [29]. Exactly these MTBs have been observed in the as-grown samples [29], and they have been demonstrated to possess many interesting properties, for example, give rise to the development of one-dimensional charge density waves [41], or be used in heterocatalysis due to their metallic states in the gap of semiconducting $MoSe_2$ sheets [42, 43].

The electronic structure of the MTBs is shown in Fig. 5. Two bands of Mo-character predominantly localized at the MTB can clearly be distinguished. The additional density of states brought in from these bands covers a wide energy range within the gap that should indeed indicate a higher conductivity along the MTBs, at

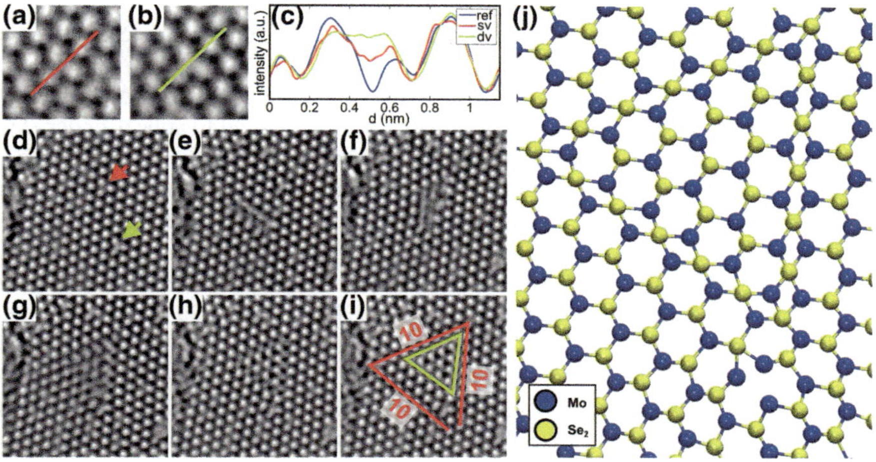

Fig. 4 Evolution of 2D $MoSe_2$ during 80 kV electron irradiation. **a** and **b** A single and double Se vacancy induced by the electron beam. **c** Line profiles measured across the single (*red line*) and double (*yellow line*) vacancies, with a line profile over pristine $MoSe_2$ lattice as a reference (*blue line*). **d** An area in $MoSe_2$, where a number of isolated vacancies are present. **e** Part of the vacancies have aggregated into a linear structure. **f** The vacancy line has reoriented itself. **g** A frame, during which many atoms within a triangular area are in motion, rendering the exact structure indiscernible. **h** A mirror-twin-domain has appeared, accompanied by an edge dislocation. **i** Frame **h**, with the *yellow lines* overlaid denoting the mirror-twin-boundary, and *red lines* with the length of 10 unit cells, indicating the presence of a dislocation. **j** The atomic structure of the newly formed mirror-twin-domain. From Ref. [29]

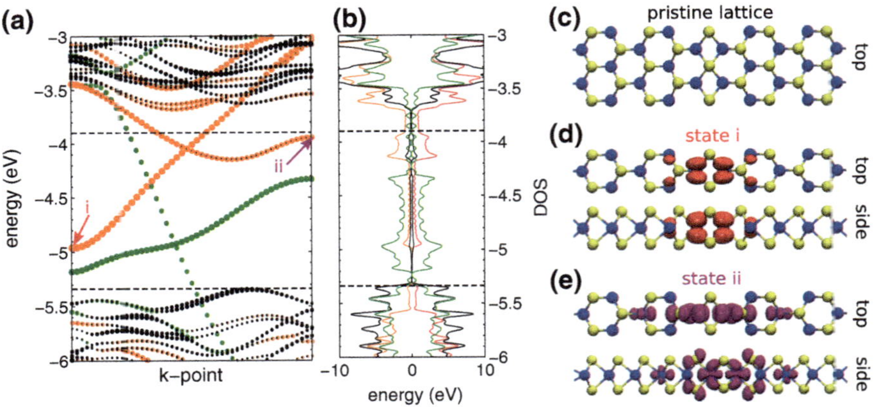

Fig. 5 Atomic and electronic structures and density of states of the MTB. **a** The band structure and **b** corresponding density-of-states from a ribbon calculation with one MTB. The states are colored by the projection to different regions of the system (*orange*: near MTB region; *green*: edge region; *black*: bulk region). **c**, **d**, **e** The atomic structure of the MTB system together with the charge density isosurface visualization of wave functions of the two MTB-localized states (*i*) and (*ii*) within the gap. From Ref. [29]

least for straight periodic boundaries, while the transmission of hole charge across the boundary was found to be suppressed [29].

In addition to the mirror-domains, and correspondingly, MTBs at their borders observed in the as-grown material, they have also been found to appear under electron beam due to vacancy coalescence [29, 38]. Specifically, the formation of mirror-domains was observed also in-situ under the beam due to production of Se vacancies under the 80 keV irradiation, Fig. 4. Due to the relatively high mass of Se, the removal of the Se atoms cannot be explained by direct knock-on damage [18], and should be attributed to inelastic processes such as ionization and electronic excitations [44] or beam-mediated chemical etching [40]. The evolution of a selected area in a $MoSe_2$ is presented in Fig. 4d–i. In frame (d), numerous isolated vacancies produced under the electron beam are already visible. During continuous exposure, the number of vacancies increases and the vacancies migrate to form a linear vacancy aggregate (frame (e)). Subsequently, the vacancy aggregate changes orientation in frame (f). In frame (g), a reorganization of the lattice inside a triangular area is underway, resulting in a locally blurred image, which makes the atomic structure indiscernible. In frame (h) the crystal has settled into a new stable configuration, i.e., a small mirror-twin-domain and a neighbouring edge dislocation (the dislocation is visible already in frame (g)), while the number of Se vacancies within the area has decreased. In panel (i), the new mirror-domain and a path of 3×10 lattice steps around the dislocation are overlaid on the micrograph of panel (h), verifying the presence of the dislocation. An atomic model of the interpreted final structure is shown in panel (j). This sequence lends further support to the hypothesis that a MTB is the energetically preferred structure in the Se-deficit limit.

Formation of an edge dislocation was shown [29] to be a prerequisite for the formation of a mirror-twin-domain inside a pre-existing $MoSe_2$ crystal when the formation path is via removal of Se atoms (as opposed to formation during the growth of the crystal): When the mirror domain is formed, the Se vacancies are eliminated, and surplus Mo is introduced at the MTBs. The extra atoms are attained by "pulling" the neighbouring crystal lattice toward the new mirror domain, which leads to the formation of the dislocation. A hole in the $MoSe_2$ crystal close to the newly created dislocation allowed the dislocation to form without the formation of a complementary oppositely oriented dislocation, which would be required in the general case of a continuous crystal.

Further insights into the formation mechanism of inversion domains in $MoSe_2$ were obtained in another in-situ study inside a scanning TEM operating at 100 keV [38]. One of the important findings was that the areas with the twin symmetry appear due to collective evolution of Se vacancies, followed by displacement of Mo atoms from their original positions.

4 Line Defects in WS_2 and WSe_2 Containing 8- and 5-Membered Rings

While vacancies in 2D MoS_2 tend to agglomerate by forming straight lines, their behavior in tungsten-containing 2D TMDs, such as WS_2 and WSe_2, proved to be completely different. A new class of point vacancy-type defects in TMDs, which can be created through 60°-rotations of metal-chalcogen bonds in the trigonal prismatic lattice of these materials was recently reported [37]. The simplest among these defects was found to be threefold symmetric trefoil-like defects, which were found abundant at elevated temperatures. An example of such defect is presented in Fig. 6b. A comparison of the experimental STEM images to the simulated ones based on possible atomic structures indicated that the system is Se-deficient due to electron beam irradiation. Similar rotational defects also exist in graphene as reconstructed divacancies, but there the C-C bond is rotated by 90° [45]. The trefoil defects were always observed in W-centered configuration, but never in the equally possible Se-centered configuration, indicating the crucial role of the Se vacancies in the formation of the defect. It should be pointed out that the actual transformation mechanism may include not only bond rotations, but also migration and rearrangements of Se atoms in the region with high vacancy concentration.

The defects, which were shown to be inherently related to the crystal symmetry of TMDs, could expand through sequential bond rotations, as revealed by the TEM experiments, and eventually form larger linear defects consisting of aligned 8-5-5-8 membered rings. The defects could also migrate through thermally activated (at elevated temperatures) and irradiation-assisted bond rotations. Eventually, the forma-

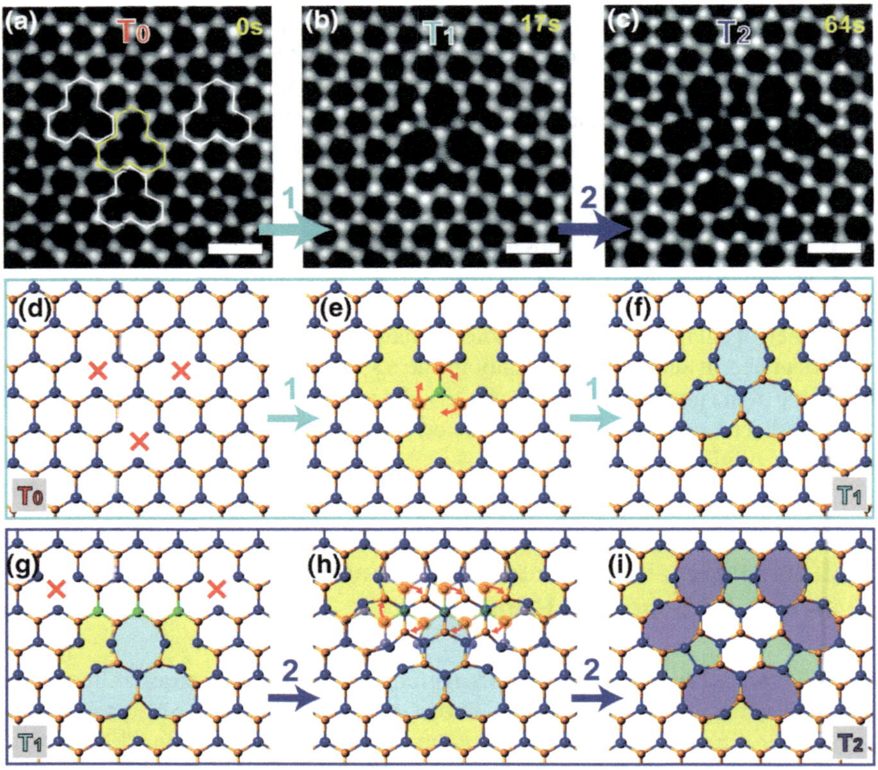

Fig. 6 Structure, formation, and evolution of rotational defect in WSe_2. The filtered annular dark field images of WSe_2 of **a** single Se vacancy (*white polygon*) and double Se vacancy (*yellow polygon*, DVSe). **b** Rotational defects. **d, f** The atomic model of the transformation including the creation of three DVSe (*red crosses*) **d** and a 60° rotation of three pairs of WSe bonds around the W atom (*green*) **e** to form a structure with three octagons (*blue*; **f**). **g, h, i** The atomic model of further transformation. **g** Two DVSe are created at the vicinity of a trefoil defect. **h** Seven pairs of WSe bond rotations according to the W atoms marked by *green* color. **i** The defect. Scale bar, 0.5 nm. From Ref. [37]

tions of inversion domains was observed, with the boundaries represented of 8-5-5-8 membered rings, Fig. 7a, d. Here the notations T_x, with $x = 0, 1, 2 \ldots$ correspond to the rank of the defect, that is related to the number of 60°-rotations of structural units in the system, see Ref. [37] for detail.

Figure 7c, e shows the electronic structure and the local density of states of such a boundary in WSe_2. The electronic states localized to the boundary have now energies in the lower mid-gap region with very small dispersion and consequently large density of states. The conductivity along these boundaries will then strongly depend on the Fermi-level position within the gap. These states were also found to carry a magnetic moment.

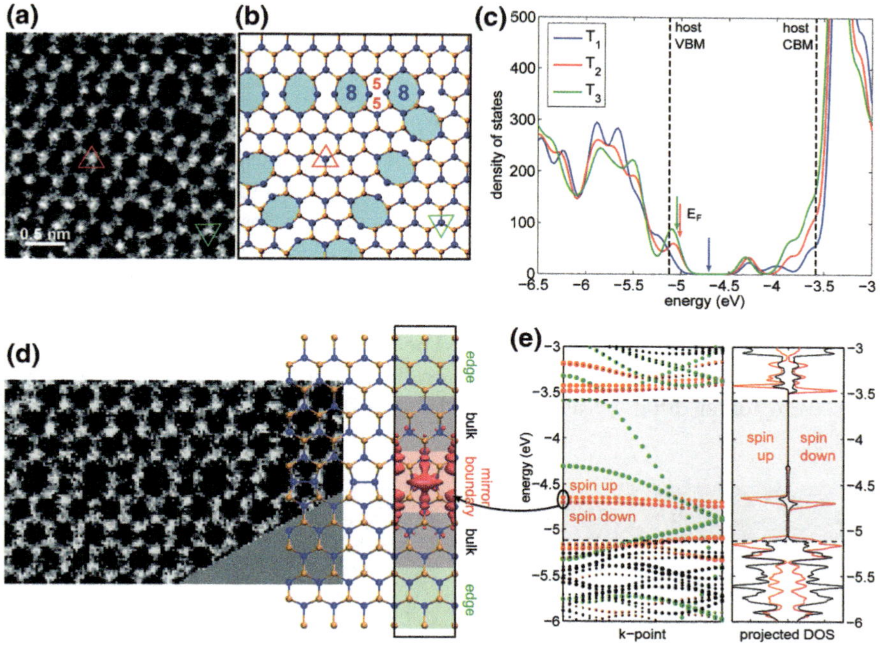

Fig. 7 Large trefoil defect and magnetic 8558 domain boundary. **a** The annular dark field image of the defect in WSe_2. Scale bar is 0.5 nm. **b** The corresponding atomic model. Larger rotational defect keep the triangular shape and include three boundaries consisting of 8558 membered rings. The orientation of the inner domain of the defect is 60° (or 180°) rotated from the outer domain. **c** Calculated density of states of the system with defects of different sizes. The Fermi-level positions are denoted with *arrows*. **d** The annular dark field image of the boundary is partially overlaid with the atomic structure. **e** The band structure and density of states of the boundary shown in (**d**) with *colors* denoting projections to different regions within the ribbon geometry used in the calculation. Edge states are shown in *green*, bulk states in *black*, and spin-polarized states associated with the boundary in *orange* and *red*. *Dashed lines* denote the band edge positions of pristine WSe_2. The wave function isosurface for the mid-gap state localized to the boundary is also shown in (**d**). Energy zero is at the vacuum. From Ref. [37]

Such rotational point and line defects were also observed in WS_2 and $MoSe_2$, but not in MoS_2, in agreement with the results of DFT calculations of energy penalties for these structures. Note also that in $MoSe_2$ MTBs shown in Fig. 4 have comparable energies as compared to those composed of 8-5-5-8 membered rings. The reason why certain line defects are preferable over other in different TMDs is still unclear and will likely be the subject of future studies.

5 Vacancy Clustering in Other TMDs

There are many other TMD materials, such, e.g., ReS_2, $NbSe_2$, and $PtSe_2$, that have recently started gaining attention [4, 46–49]. However, very little is still known about vacancies, vacancy clustering, and grain boundaries in these materials. This can be partly ascribed to the sensitivity of these materials to the electron irradiation and also to the processing steps required to produce the samples.

Monolayer ReS_2 and $ReSe_2$ were studied in Ref. [46]. The atomic structure of these materials shows a clear "diamond-chain" pattern, as depicted in Fig. 8a, b. Upon electron irradiation the vacancies are seen to be located primarily in the trenches between the diamond chains. The inequivalent positions for the vacancies are depicted in Fig. 8b. Depending on the position, calculated formation energies were found differ by as much as 1.2 and 0.85 eV in the case of ReS_2 and

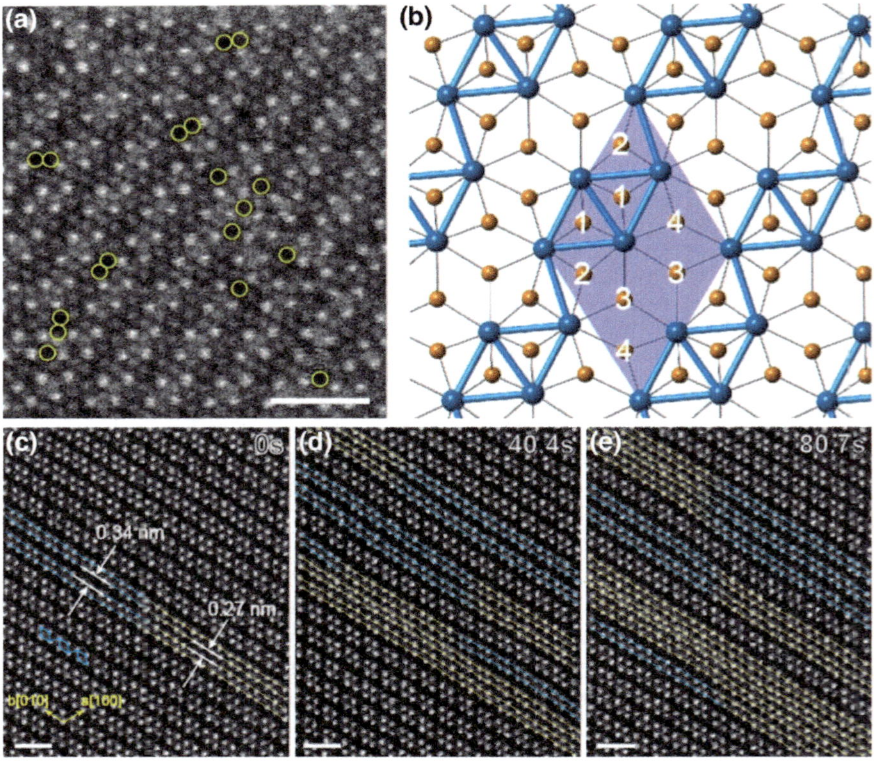

Fig. 8 **a** The annular dark field image of $ReSe_2$. *Yellow circles* indicate the Se monovacancies. Scale bar is 1 nm. **b** Illustration of the atomic structure (*blue*: Re, *Yellow*: chalcogen) and the enumeration of the nonequivalent chalcogen atoms. **c**, **d**, **e** Re diamond chains in ReS_2 zip up progressively in the b direction. The original diamond chain spacing is 0.34 nm marked in *blue lines*. The zipped spacing between diamond chains is 0.27 nm marked in *yellow lines*. From Ref. [46]

$ReSe_2$, respectively. In agreement with the experiment, the lowest formation energies were obtained for vacancies in the trenches, where the vacancies are then expected to migrate. Upon continued irradiation, the atomic structures of ReS_2 were found to undergo distinct change where the diamond chains are "zipped" together (cf. Fig. 8c, d, e), which is mitigated by the aggregation of vacancies in sufficiently high concentration to the trenches.

Unlike all the other considered TMD materials, vacancies in $NbSe_2$ do not aggregate. Instead, vacancies in the nearest neighbor sites, whether in the same or different side of the sheet, feel effectively a repulsive interaction [50].

6 Conclusions and Outlook

To conclude, numerous in-situ TEM experiments demonstrated that point defects in TMDs induced by electron beam irradiation can coalesce into line defects, but their quasi-one-dimensional atomic structure varies from member to member in the transition metal dichalcogenides family. The coalescence mechanisms involve thermally activated and beam-induced migration of vacancies, as well as instabilities in the atomic network when vacancy concentration increases. The agglomeration of vacancies in TMDs can give rise to formation of grains embedded into a larger grain with a different orientation of the crystal lattice of the material. Depending on the chemical content of TMDs, line defects with different morphologies are formed. For example, agglomeration of vacancies in MoS_2 gives rise to straight vacancy lines without any substantial local transformations of the atomic structure, while the sputtering of chalcogen atoms in WSe_2 and WS_2 results in the development of rotational defects and eventually mirror grains with the boundaries consisting of 5- and 8-member rings. In $MoSe_2$ the appearance of mirror grains is accompanied with the development of different line structures at the borders. The reason why certain line defects are preferable over other in different TMDs is still unclear and requires further studies.

The line defect structures have many interesting characteristics and affect mechanical, electronic and optical properties of TMDs. As their orientation depends on strain (and also likely local charging), electron irradiation can be a tool to engineer the properties of 2D TMDs. It should also be pointed out that many 2D TMDs exist in two different phases: the H (trigonal prismatic D_{3h}) and T (octahedral O_h) phases with different electronic properties. Further reconstructions of the T phases are possible, and the borders between the phases can be also referred to as line defects [51], so that understanding the formation and evolution of line defects under electron beam can also help to induce local phase transformations in 2D TMDs.

Acknowledgements The Authors would like to thank for many years of successful collaboration our coworkers, whose contribution to the findings presented in this chapter cannot be overestimated: Ute Kaiser, Kazu Suenaga, Yung-Chang Lin, Ossi Lehtinen, Torbjörn Björkman, Artem Pulkin, Tibor Lehnert, Oleg V. Yazyev, Andras Kis, and Simon Kurasch.

References

1. Novoselov, K.S., Geim, A.K., Morozov, S.V., Jiang, D., Zhang, Y., Dubonos, S.V., Grigorieva, I.V., Firsov, A.A.: Science **306**, 666 (2004)
2. Bae, S., Kim, H., Lee, Y., Xu, X., Park, J., Zheng, Y., Balakrishnan, J., Lei, T., Kim, H.R., Song, Y.I., Kim, Y.J., Kim, K.S., Özyilmaz, B., Ahn, J.H., Hong, B.H., Iijima, S.: Nat. Nanotech. **5**, 574 (2010)
3. Wang, Q.H., Kalantar-Zadeh, K., Kis, A., Coleman, J.N., Strano, M.S.: Nat. Nanotech. **7**, 699 (2012)
4. Nicolosi, V., Chhowalla, M., Kanatzidis, M.G., Strano, M.S., Coleman, J.N.: Science **340**, 1420 (2013)
5. Chhowalla, M., Shin, H.S., Eda, G., Li, L.J., Loh, K.P., Zhang, H.: Nat. Chem. **5**, 263 (2013)
6. Geim, A.K., Grigorieva, I.V.: Nature **499**(7459), 419 (2013). doi:10.1038/nature12385
7. Mak, K.F., Shan, J.: Nat. Photonics **10**, 216 (2016)
8. Muller, D.A.: Nat. Mat. **8**(4), 263 (2009)
9. Sawada, H., Sasaki, T., Hosokawa, F., Suenaga, K.: Phys. Rev. Lett. **114**, 166102 (2015)
10. Krivanek, O.L., Chisholm, M.F., Nicolosi, V., Pennycook, T.J., Corbin, G.J., Dellby, N., Murfitt, M.F., Own, C.S., Szilagyi, Z.S., Oxley, M.P., Pantelides, S.T., Pennycook, S.: Nature **464**, 571 (2010)
11. Kepaptsoglou, D., Hardcastle, T.P., Seabourne, C.R., Bangert, U., Zan, R., Amani, J.A., Hofsäss, H., Nicholls, R.J., Brydson, R.M.D., Scott, A.J., Ramasse, Q.M.: ACS Nano **9**, 11398 (2015)
12. Suenaga, K., Koshino, M.: Nature **468**, 1088 (2010)
13. Lin, Y.C., Teng, P.Y., Chiu, P.W., Suenaga, K.: Phys. Rev. Lett. **115**, 206803 (2015)
14. Krivanek, O.L., Lovejoy, T.C., Dellby, N., Aoki, T., Carpenter, R.W., Rez, P., Soignard, E., Zhu, J., Batson, P.E., Lagos, M.J., Egerton, R.F., Crozier, P.A.: Nature **514**, 209 (2014)
15. Lehtinen, O., Kurasch, S., Krasheninnikov, A.V., Kaiser, U.: Nat. Commun. **4**, 2098 (2013)
16. Robertson, A.W., Allen, C.S., Wu, Y.A., He, K., Olivier, J., Neethling, J., Kirkland, A.I., Warner, J.H.: Nat. Commun. **3**, 1144 (2012)
17. Kotakoski, J., Krasheninnikov, A.V., Kaiser, U., Meyer, J.C.: Phys. Rev. Lett. **106**, 105505 (2011)
18. Komsa, H.P., Kotakoski, J., Kurasch, S., Lehtinen, O., Kaiser, U., Krasheninnikov, A.V.: Phys. Rev. Lett. **109**, 035503 (2012)
19. Zhou, W., Zou, X., Najmaei, S., Liu, Z., Shi, Y., Kong, J., Lou, J., Ajayan, P.M., Yakobson, B.I., Idrobo, J.C.: Nano Lett. **13**, 2615 (2013)
20. Hong, J., Hu, Z., Probert, M., Li, K., Lv, D., Yang, X., Gu, L., Mao, N., Feng, Q., Xie, L., Zhang, J., Wu, D., Zhang, Z., Jin, C., Ji, W., Zhang, X., Yuan, J., Zhang, Z.: Nat. Commun. **6**, 6293 (2015)
21. Lin, Y.C., Dumcenco, D.O., Komsa, H.P., Niimi, Y., Krasheninnikov, A.V., Huang, Y.S., Suenaga, K.: Adv. Mat. **26**, 2857 (2014)
22. Gibb, A.L., Alem, N., Chen, J.H., Erickson, K.J., Ciston, J., Gautam, A., Linck, M., Zettl, A.: J. Am. Chem. Soc. **135**, 6758 (2013)
23. Björkman, T., Kurasch, S., Lehtinen, O., Kotakoski, J., Yazyev, O.V., Srivastava, A., Skakalova, V., Smet, J.H., Kaiser, U., Krasheninnikov, A.V.: Scientific Rep. **3**, 3482 (2013)
24. Yazyev, O.V., Chen, Y.P.: Nat. Nanotech. **9**, 755 (2014)
25. Huang, P.Y., Ruiz-Vargas, C.S., van der Zande, A.M., Whitney, W.S., Levendorf, M.P., Kevek, J.W., Garg, S., Alden, J., Hustedt, C.J., Zhu, Y., Park, J., McEuen, P.L., Muller, D.A.: Nature **469**, 389 (2011)
26. Azizi, A., Zou, X., Ercius, P., Zhang, Z., Elías, A.L., Perea-López, N., Stone, G., Terrones, M., Yakobson, B.I., Alem, N.: Nat. Commun. **5**, 4867 (2014)
27. Najmaei, S., Liu, Z., Zhou, W., Zou, X., Shi, G., Lei, S., Yakobson, B., Idrobo, J.C., Ajayan, P.M., Lou, J.: Nat. Mat. **12**, 754 (2013)
28. van der Zande, A.M., Huang, P.Y., Chenet, D.A., Berkelbach, T.C., You, Y., Lee, G.H., Heinz, T.F., Reichman, D.R., Muller, D.A., Hone, J.: Nat. Mat. **12**, 554 (2013)

29. Lehtinen, O., Komsa, H.P., Pulkin, A., Whitwick, M.B., Chen, M.W., Lehnert, T., Mohn, M.J., Yazyev, O.V., Kis, A., Kaiser, U., Krasheninnikov, A.V.: ACS Nano **9**, 3274 (2015)
30. Yong, K.S., Otalvaro, D.M., Duchemin, I., Saeys, M., Joachim, C.: Phys. Rev. B **77**, 205429 (2008)
31. Gibertini, M., Marzari, N.: Nano Lett. **15**(9), 6229 (2015)
32. Le, D., Rahman, T.S.: J. Phys. Conden. Matter **25**, 312201 (2013)
33. Ly, T.H., Perello, D.J., Zhao, J., Deng, Q., Kim, H., Han, G.H., Chae, S.H., Jeong, H.Y., Lee, Y.H.: Nat. Commun. **7**, 10426 (2016)
34. Wang, S., Lee, G.D., Lee, S., Yoon, E., Warner, J.H.: ACS Nano p. acsnano.6b01673 (2016). doi:10.1021/acsnano.6b01673
35. Zhang, Z., Zou, X., Crespi, V.H., Yakobson, B.: ACS Nano **7**, 10475 (2013)
36. Komsa, H.P., Kurasch, S., Lehtinen, O., Kaiser, U., Krasheninnikov, A.V.: Phys. Rev. B **88**, 035301 (2013)
37. Lin, Y.C., Björkman, T., Komsa, H.P., Teng, P.Y., Yeh, C.H., Huang, F.S., Lin, K.H., Jadczak, J., Huang, Y.S., Chiu, P.W., Krasheninnikov, A.V., Suenaga, K.: Nat. Commun. **6**, 6736 (2015)
38. Lin, J., Pantelides, S.T., Zhou, W.: ACS Nano **9**, 5189 (2015)
39. Han, Y., Hu, T., Li, R., Zhou, J., Dong, J.: Phys. Chem. Chem. Phys. **17**, 3813 (2015)
40. Meyer, J.C., Eder, F., Kurasch, S., Skakalova, V., Kotakoski, J., Park, H., Roth, S., Chuvilin, A., Eyhusen, S., Benner, G., Krasheninnikov, A.V., Kaiser, U.: Phys. Rev. Lett. **108**, 196102 (2012)
41. Barja, S., Wickenburg, S., Liu, Z.F., Zhang, Y., Ryu, H., Ugeda, M.M., Hussain, Z., Shen, Z.X., Mo, S.k., Wong, E., Salmeron, M.B., Wang, F., Crommie, M.F., Ogletree, D.F., Neaton, J.B., Weber-Bargioni, A.: Nat. Phys. **X**, 1 (2016). doi:10.1038/nphys3730
42. Liu, H., Jiao, L., Yang, F., Cai, Y., Wu, X., Ho, W., Gao, C., Jia, J., Wang, N., Fan, H., Yao, W., Xie, M.: Phys. Rev. Lett. **113**, 066105 (2014)
43. Liu, H., Zheng, H., Yang, F., Jiao, L., Chen, J., Ho, W., Gao, C., Jia, J., Xie, M.: ACS Nano **9**, 6619 (2015)
44. Zan, R., Ramasse, Q., Jalil, R., Georgiou, T., Bangert, U., Novoselov, K.S.: ACS Nano **7**, 10167 (2013)
45. Banhart, F., Kotakoski, J., Krasheninnikov, A.V.: ACS Nano **5**, 26 (2011)
46. Lin, Y.C., Komsa, H.P., Yeh, C.H., Bjrkman, T., Liang, Z.Y., Ho, C.H., Huang, Y.S., Chiu, P.W., Krasheninnikov, A.V., Suenaga, K.: ACS Nano **9**(11), 11249 (2015)
47. Tongay, S., Sahin, H., Ko, C., Luce, A., Fan, W., Liu, K., Zhou, J., Huang, Y.S., Ho, C.H., Yan, J., Ogletree, D.F., Aloni, S., Ji, J., Li, S., Li, J., Peeters, F.M., Wu, J.: Nat. Commun. **5**, 3252 (2014)
48. Cao, Y., Mishchenko, A., Yu, G.L., Khestanova, E., Rooney, A.P., Prestat, E., Kretinin, A.V., Blake, P., Shalom, M.B., Woods, C., Chapman, J., Balakrishnan, G., Grigorieva, I.V., Novoselov, K.S., Piot, B.A., Potemski, M., Watanabe, K., Taniguchi, T., Haigh, S.J., Geim, A.K., Gorbachev, R.V.: Nano Lett. **15**, 4914 (2015)
49. Wang, Y., Li, L., Yao, W., Song, S., Sun, J.T., Pan, J., Ren, X., Li, C., Okunishi, E., Wang, Y.Q., Wang, E., Shao, Y., Zhang, Y.Y., Yang, H.T., Schwier, E.F., Iwasawa, H., Shimada, K., Taniguchi, M., Cheng, Z., Zhou, S., Du, S., Pennycook, S.J., Pantelides, S.T., Gao, H.J.: Nano Lett. **15**, 4013 (2015)
50. H.-P. Komsa et al.: Unpublished
51. Lin, Y.C., Dumcenco, D.O., Huang, Y.S., Suenaga, K.: Nat. Nanotech. **9**, 391 (2014)

Morphing Graphene-Based Systems for Applications: Perspectives from Simulations

T. Cavallucci, K. Kakhiani, R. Farchioni and V. Tozzini

Abstract Graphene, the one-atom-thick sp^2-hybridized carbon crystal, displays unique electronic, structural and mechanical properties, which promise a large number of interesting applications in diverse high-tech fields. Many of these applications require its functionalization, e.g., with substitution of carbon atoms or adhesion of chemical species, creation of defects, modification of structure or morphology, to open an electronic band gap to use it in electronics, or to create 3D frameworks for volumetric applications. Understanding the morphology–properties relationship is the first step to efficiently functionalize graphene. Therefore, a great theoretical effort has been recently devoted to model graphene in different conditions and with different approaches involving different levels of accuracy and resolution. Here, we review the modeling approaches to graphene systems, with a special focus on atomistic level methods, but extending our analysis onto coarser scales. We illustrate the methods by means of applications with possible potential impact.

1 Introduction

Since its first isolation [1] and especially after the assignment of the Nobel Prize for Physics in 2010 for its discovery, graphene has triggered great expectations. Most of its properties depend on the coincidence of a number of physical–chemical circumstances: three of the four valence electrons of carbon organize in three sp^2-hybridized orbitals producing an exactly planar geometry, while the forth contributes to a vast electronic delocalization over the plane, giving great stability to

T. Cavallucci · K. Kakhiani · V. Tozzini (✉)
NEST-Scuola Normale Superiore, Pisa, Italy
e-mail: valentina.tozzini@nano.cnr.it

T. Cavallucci · K. Kakhiani · R. Farchioni · V. Tozzini
Istituto Nanoscienze del Cnr, Piazza San Silvestro 12, 56127 Pisa, Italy

R. Farchioni
Dipartimento di Fisica, Università di Pisa, Largo Bruno Pontecorvo, 56127 Pisa, Italy

V. Morandi and L. Ottaviano (eds.), *GraphITA*, Carbon Nanostructures,
DOI 10.1007/978-3-319-58134-7_7

the electronic and geometric structure of the single sheet and very weak interlayer interaction mediated only by van der Waals (vdW) forces and favoring bidimensionality. In addition, the specific geometry of in-plane bonds generates the highly symmetric honeycomb lattice, responsible for the peculiar band structure with conic geometry at the K points of the Brillouin Zone (BZ). This is responsible for the exotic electronic properties of graphene [2], such as the ultra-relativistic-like high mobility conduction and the possible existence of Majorana fermions [3].

Its—equally exceptional—mechanical properties, have received less attention. The peculiar chemistry and symmetry also brings an extremely large resistance to tensile strain especially considering its bidimensionality (Young modulus of the order of TPa [4], fivefold that of steel), coupled to very low out-of-plane (bending) rigidity κ (1–2 eV [5–7]) which is at least one order of magnitude smaller than that expected from materials with comparable in-plane rigidity [8]. This implies an extreme flexibility associated to strength suggesting a vast range of applications in a variety of high-tech fields [9], including those involving exposure to extreme environmental conditions, such as aero-space technologies [10]. Flexibility and bi-dimensionality also imply the existence of low-frequency quadratically dispersive acoustic phonon branches associate to out-of-plane displacement ("flexural" phonons), which can be described as traveling ripples [11] and have a main role in thermal behavior and related phenomena [12].

Many of the applications, however, require graphene morphology manipulation of some kind (see Fig. 1). Considering for instance nanoelectronics, graphene has null density of carriers at the Fermi level, therefore needing doping to be used as conductor, or band gap opening to be used as semiconductor. These properties can be obtained by escaping from the perfect infinite 2D crystal case, e.g., in multilayers or breaking the honeycomb lattice symmetry. Graphene symmetry disruption is not difficult per se, but it is difficult to achieve in a controlled fashion.

Graphene is lightweight and with a huge surface-to-mass ratio, therefore is considered potentially interesting for gas storage applications. However, it is rather inert, little reactive and with weak vdW interactions. Therefore, this class of applications requires enhancing either physical or chemical interactions. Considering for instance hydrogen storage [13] for energy applications, accurate experimental evaluations of physisorption [14] show that a limiting value for gravimetric (i.e., the relative hydrogen mass stored with respect to the mass system) density in graphene-based systems is $\sim$1% at room temperature (5–6% at cryogenic temperatures). On the other hand, considering chemisorption one can in principle reach 6–8% gravimetric density at room temperature, but high chemi(de)sorption barriers makes the kinetic of the process very slow at room temperature [13]. Therefore, efforts are directed both to enhance physical-like interactions and to lower the kinetic barriers for chemisorption. Again, manipulation of morphology is required. As for electronics, achievement of interesting properties must proceed through rupture of the perfect graphene symmetry.

While for electronics 2-dimensionality is useful, for other applications the large surface–to-mass ratio must be declined in a 3D context. This is the case of energy storage applications, including also batteries and supercapacitors, besides hydrogen

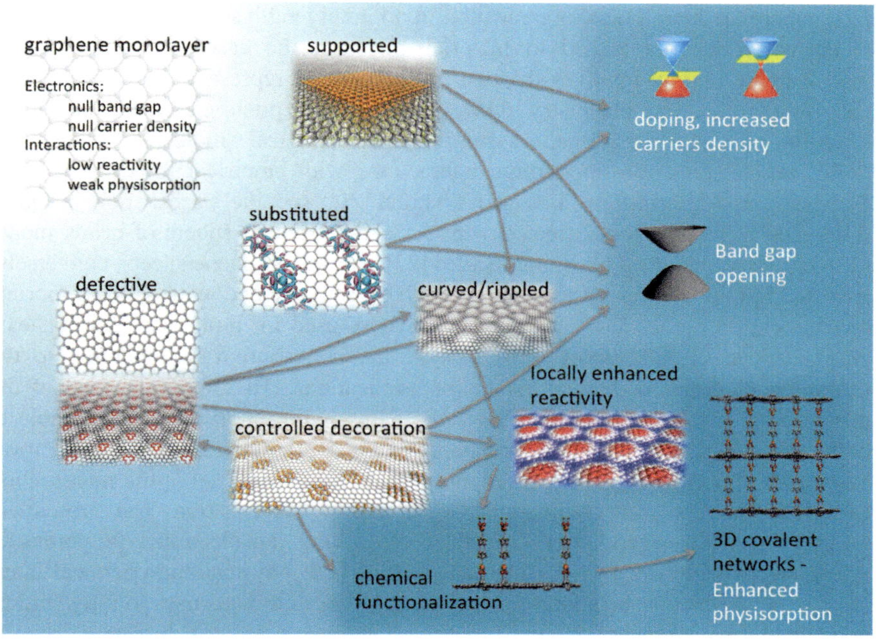

Fig. 1 A schematic illustration of the relationship between graphene modification and properties in graphene. The bare single-layer graphene is located in the upper left corner. Background increasing level of shade is roughly proportional to the level of manipulation, from simpler ones (inclusion of defects or substitutions, rippling) to the creation of covalent networks. The effects on electronic and chemical properties and physical interactions are reported in white. The *arrows* indicate the interrelationships: e.g., the substrate creates rippling and in some cases doping and gap opening; chemical substitutions with N and B induce chemical doping and bad gap opening; defects open the gap and induce curvature; curvature in turn induces local change in reactivity, and consequently controlled decoration and band gap opening. Finally, locally enhanced reactivity could be used to control decoration, chemical functionalization, and finally to build 3D covalent network with enhanced physisorption properties. (The images of the chemical functionalization and 3D networks were adapted from Ref. [16])

(or other gases) storage. Therefore great efforts are directed to building 3D graphene-based frameworks [15]. This involves including spacers between sheets [16] with tailored size and mechanical properties, and control their amount and distribution [17]. Building 3D networks, controlled chemical functionalization is crucial in a number of other applications [18], including biomedical ones [19].

It appears that manipulation of graphene morphology, structure, electronic properties, chemistry, and functionalization are interconnected tasks assuming a key role in all emerging graphene-related technologies (see Fig. 1). In this context, theoretical investigations and computer modeling are tools of outmost importance to study and possibly design properties of these new materials. The so-called multi-scale approach, traditionally designed and used for biomolecular systems

[20], where the hierarchical organization in different length scales is more apparent, is currently being extended to materials [21]. In the cases under discussion, multi-scaling mainly involves the two basic levels of representations, namely the "ab initio" quantum chemistry (QM) level, with explicit electrons capable of describing the chemistry of the system, and the classical "molecular mechanics" (MM) representing the interatomic interactions with empirical force fields (FF). Super-atomic descriptions are also "natural" in specific cases: For instance, fullerenes were sometimes treated with a Coarse Grained sphere-of-beads model [22] similar to that used for biomolecules [23]. For the graphene sheets, conversely, a natural low-resolution representation is a 2D membrane endowed with appropriate mechanical properties [24], analogously to those used for biological membranes.

The scope of this report is to analyze the contribution of modeling to the morphing of graphene-based system for applications. Therefore, for each of the different types of morphing described, the focus is on theory and models able to better describe and predict the properties under examination. The well-known properties of perfect single-sheet graphene are also summarized in the next section to have a base for comparison for changes due to manipulation. The subsequent section describes several ways aimed at controlling the electronic properties in graphene-derived systems. The manipulation of both physical and chemical interactions of graphene is addressed in Sect. 4. Section 5 focuses on dynamical deformation of graphene. Section 6 reports a summary and conclusions.

2 Single-Sheet Suspended Graphene

The electronic and structural properties of graphene are very well known and described in a number of very good reviews [2]. The minimal unit cell of graphene (represented in blue in Fig. 2a) includes two C atoms. The eight valence electrons organize in four filled bands, whose structure was calculated with all available electronic structure methods [25]. The band structure evaluated within the Density Functional Theory (DFT) framework (PBE functional [26]) is reported in Fig. 2c (left plot). The π band crosses its empty counterpart (in red, the filled bands are in blue) at the K point, where the linear dispersion is described in terms of the Fermi velocity v_F. Inclusion of many body effects show on an average relatively small corrections to the DFT picture, resulting in a renormalization of v_F, whose entity is still under debate [27].

Clearly, when treating an isolated single-sheet graphene with no defects or structural deformation, the unit cell is usually the preferred choice for the model system. However, there are infinite ways of defining a graphene supercell. Larger, rotated, or differently shaped cells might be necessary in the presence of substrates or other elements breaking the symmetry of the honeycomb lattice. A selection of them is reported in Fig. 2a, while their corresponding Brillouin Zones (BZ) in Fig. 2b. The size, shape, and orientation of the BZs change accordingly. The definition of different BZs implies that the description of the band structure is cell

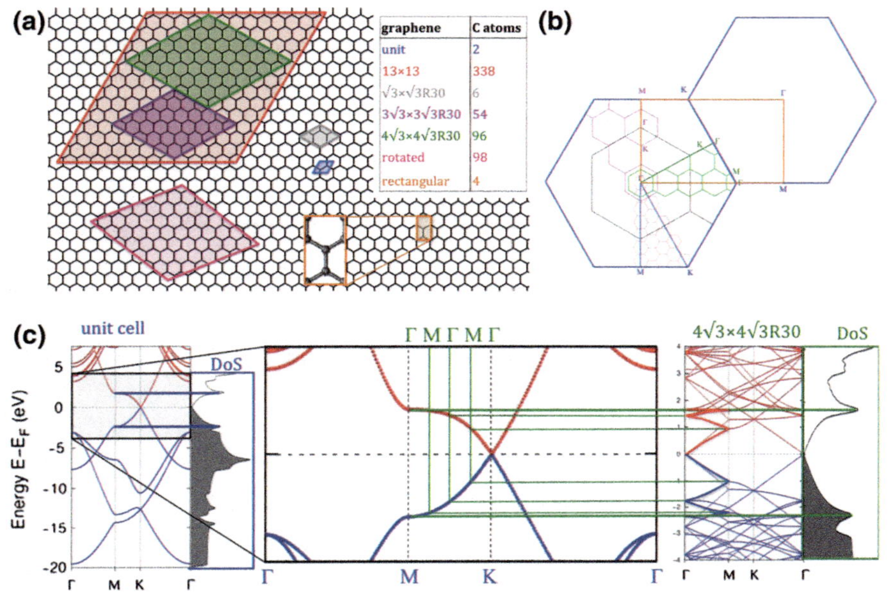

Fig. 2 Equivalent representation of the graphene structure. **a** Representation of graphene cells of different sizes, orientation, and shape. The number of atoms included is reported. **b** Representation of the BZ of the cells reported in (**a**) (except the one rotated with 98 atoms) with the main symmetry points. Color coding is the same as in (**a**) and the symmetry points are colored as the cell BZ they refer to. **c** Electronic structure of single layer flat graphene evaluated within DFT-PBE approach (see text). On the left: Band structure and DoS for the unit cell [in *blue* in (**a**)]; Center: same expanded in a −4 eV–+4 eV. When the $4\sqrt{3} \times 4\sqrt{3}$R30 cell is used, the M-K line of unit cell is fragmented in the Γ-M-Γ-M-Γ line [see (**b**)], as represented by green lines. This produces a refolding of the band, which is represented in the band structure of $4\sqrt{3} \times 4\sqrt{3}$R30 reported on the right (with its DoS). Filled bands are in *blue*, empty in *red*. In the DoS, filled states are in *grey*, empty in *white*. The edges of the π bands are highlighted with *horizontal thick lines* (*blue* in the unit cell bands, *green* in the $4\sqrt{3} \times 4\sqrt{3}$R30 bands structure)

dependent. As an example, in Fig. 2c we report the comparison of the graphene band structure evaluated in DFT-PBE in unit cell and in the $4\sqrt{3} \times 4\sqrt{3}$R30 cell [26, 28] (colored in green in Fig. 2a, b), including 96 C atoms. Once the scale of energies of the two band structures are aligned, the bands of the $4\sqrt{3} \times 4\sqrt{3}$R30 cell appear refolded. In order to understand the refolding, we follow for instance the band along the M-K line of the unit cell, which is fragmented in a sequence of Γ-M-Γ-M-Γ lines of the $4\sqrt{3} \times 4\sqrt{3}$R30 cell, as it can be seen in Fig. 2b. The fragmentation is reported in the central plot, and the refolded empty and filled π bands are highlighted in the right band structure. Interestingly, the crossing of bands at K point is remapped onto the Γ point in the $4\sqrt{3} \times 4\sqrt{3}$R30 cell, as an effect of refolding/rotation. It is also interesting to remark that, while the band representation changes, the value of physical observables should not. As a matter of fact the density of electronic states (DoS) evaluated in the two cases coincide (within numerical error) once the energy scales are aligned. This is true in particular for the two peaks at ±1.5 eV appearing where the π bands become flat (π edges).

Finally, we remark that the use of rectangular cells allows remapping the main symmetry directions along the two Cartesian directions x and y.

3 Modification of Graphene Electronic Properties

Any breaking of the graphene perfect symmetry induces changes in the electronic or transport properties. In the following we review the effects of substrates, chemical substitution/adhesion and structural modifications on the electronic properties.

3.1 B and N Substitutions and Structural Defects

Hexagonal Boron Nitride, BN, has a 2D honeycomb structure identical with that of graphene, with the two triangular sublattices occupied by B and N, respectively. The sublattice symmetry breaking induces the opening of a large band gap, estimated in ~4.5 eV within the DFT scheme (generalized gradient approximation [29]), making it the insulating material more similar to graphene. Graphene partially substituted with BN patches or strips of different configuration shows intermediate behavior and can be considered therefore semiconductor system with band gap tunable as a function of the amount and size of BN areas with respect to graphene. Independent DFT estimates [26, 29, 30] indicate that the band gap opening is roughly linear at low B/N density, and can be estimated in ~0.03–0.06 eV for 1% of substituted C atoms. In presence of excess of N or B, doping (of n-type or p-type respectively) is also present, estimated in ±0.3–0.4 eV of Fermi level shift with respect to the Dirac point for each % point of N(B) excess [31]. In this case, for given shapes of the substitution patches, mid-gap states and specific magnetic properties can also appear.

Purely structural defects in graphene such as isolated or aligned dislocations delimiting grain boundaries can also produce severe changes in the electronic structure [32]. Depending on the specific periodicity and topology, and on the relative orientation of domains, the system can display conductive behavior or band gaps with reflective transport behavior at the grain boundaries [33]. Vacancy type and Stone–Wales type defects can open band gaps on the scale of eV leading to semiconducting behavior [34] which opens to the possibility of using defects to engineer the electronic properties of graphene. Clearly, due to rupture of the regular π electronic system, structural defects of any kind also produce structural deformation and changes in local reactivity, which are discussed in Sect. 4.

3.2 Effects of the Substrate

In supported graphene, the interaction with the substrate can induce electronic properties modifications directly, or indirectly (e.g., by substrate-induced defects) or influence electronic modification due to other effects. Considering graphene grown on SiC by Si evaporation, for instance, it was early inferred by DFT studies that—for grown on the C-rich face—the substrate can enhance the effect of Al, P, N, and B substitutional doping [35]. Of all dopants, N is shown to prefer substituting the graphene upper layer, while other dopants prefer interstitial or buffer layer locations. The enhancement effect seems due to an increased stability of dopants operated by interaction with the substrate. In a model without the buffer layer [36] (whose presence is still debated when grown on the C-exposed surface) a direct effect of the substrate producing n-type doping of ~0.3–0.4 eV, was reported, very sensitive to the intercalation of passivant H atoms. This work also reports a marked sensitivity to the use of different density functionals and treatments of vdW interactions, confirming a main role of the interaction with the substrate. In the Si-exposed surface the presence of a covalently bond buffer layer was early established, and calculations with a minimalist cell show noticeable levels of doping [37].

Early calculations however, were performed in very small simulation cells, inducing strain on the graphene layers to achieve commensuration with the substrate. Subsequent calculations in larger and more relaxed supercells revealed a reduction of all the effects: on the C-rich surface, the first graphene layer becomes almost detached and recovers graphene-type metallicity, with little doping [38]. In the case of the Si-rich surface the stress can be completely relaxed only using a very large supercell [39], where doping seems to be negligible. The question is still controversial however, because experimental observations indicate a variable level of n-doping in monolayer graphene, which tends to decrease in quasi-free standing graphene (QFSG) obtained intercalating H underneath the buffer layer, depending on the amount of H, and negligible interactions of QFSG with the substrate are confirmed by DFT calculations in the fully relaxed model [40], reporting a ~5–8 meV band gap and basically no doping [41]. A more advanced theoretical analysis indicates that perturbation from the noninteracting situation and others effects (such as the appearance of specific non linearities in the bands) might be due to many body electron interactions and phonon-coupling effects not included in the standard DFT treatment [42]. Conversely, metallic substrates induce a genuine doping whose origin is ascribed to the difference in the work function and to the more or less "physical" or "chemical" interactions between graphene-metal. A systematic DFT based study shows a n-type doping for Al, Ag, Cu, and p-type for Au and Pd [43].

Effects of substrates and of defects can combine. For instance, vacancy-type defects created on QFSG are shown to strongly interact with the H coverage and induce doping and/or localized states [44] and metals intercalation (e.g., Li) produces a strong n-type doping. In addition, the procedures to induce substitutional doping, e.g. with N, often induce also vacancy type nitrogenated defects [45] (e.g., with pyridinic or pyrrolic type reconstruction), which combines and often enhances the doping and band gap opening effects, allowing in principle a fine tuning of the electronic properties of the graphene sheet [46].

3.3 Effects of Adatoms

Chemisorption of adatoms (typically H and F) is another common way to change graphene properties. Because the fully hydrogenated sheet—graphane—is an insulator [47], partially hydrogenated graphene display semiconducting properties, with band gap depending on the coverage and decoration. Decoration in strips produces the typical ribbon-like behavior [48], with gap dependent on the edge type (zigzag or armchair) and decreasing with the width of the hydrogenated strips, and therefore increasing with the H coverage. Decoration in patches produces very variable situations [49, 50] depending on the shape and connectivity of the patches, sometimes including mid-gap nondispersive states [51]. However, when experimental and DFT data are put together as a function of the H coverage, irrespective of the decoration type (Fig. 3), they seem to accumulate onto a line which was empirically fitted with an almost square root behavior (precisely, a 0.6 exponential [51]). Best adherence to the empirical curve is observed for the uniform coverage [52] or in general in experimental cases, when the coverage is likely to be more random. Cases of regular or symmetric coverage seem to bring larger dispersion from the curve. Considering the dispersion of the curve, a measurement of the band gap, e.g., with STS techniques could give an evaluation of H coverage with an error of 10–15%.

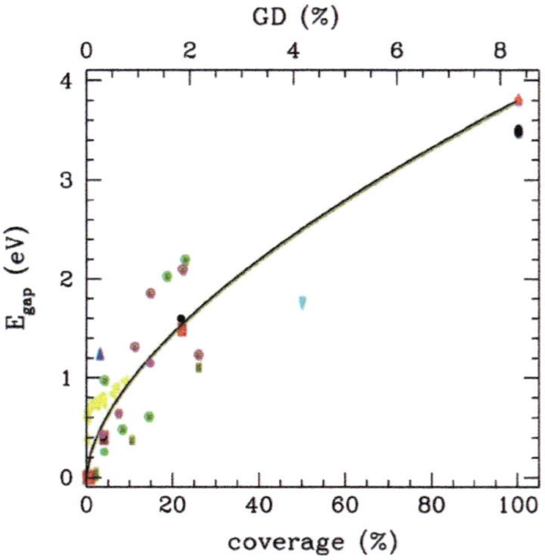

Fig. 3 Band with of partially hydrogenated graphene as a function of the H coverage. Data coding: *line* and *magenta/green/brown circles/dots* are taken from Ref. [48] (DFT); *black dots* are from Ref. [45] (DFT); *red squares* are from Ref. [47] (expt and DFT); *yellow* data are from Ref. [50] (DFT and expt); *blue* and *cyan triangles* from Ref. [49] (DFT)

Due to its symmetry, single-layer graphene is not piezoelectric, although there are indications of its reactivity to electric fields by means of flexoelectric behavior [53, 54]. Again, the sheets functionalization with specific elements (H, F, Li, K) was shown an effective way to induce piezoelectricity: fluorination and decoration with Lithium, in particular, are shown to bring piezoelectric coefficients of the order of magnitude of those of 3D materials [55]. Piezoelectricity can also be induced creating holes with the right symmetry on the layer [56], again breaking the graphene inversion symmetry. Chemical manipulation and substitution also influences the flexoelectric properties: curvature changes induced by an external electrostatic field are enhanced in the presence of N substitutions [26] (Fig. 4).

Table 1 reports a summary of the cited DFT-based works on graphene with different morphological modifications.

In conclusion to this section, we remark that while standard DFT methods seem appropriate to evaluate the energetics and electronic properties for chemical manipulations, the effect of substrates, especially when the interaction is more physical in nature, calls into play the use of more advanced DFT schemes, involving the use of vdW corrections and evaluation of many body electronic effects.

4 Reactivity and Interactions Manipulation

From previous discussion the necessity of controlling adsorption or substitution of chemical species onto graphene emerges as a way to endow graphene with electronic and electromechanical properties of outmost interest in applications. Controlling interaction of graphene with atomic or molecular species has also a direct use in energy applications: graphene has a huge surface-to-mass ratio, therefore it would be in principle an ideal candidate not only for gas (mainly hydrogen) storage applications, but also for supercapacitors and batteries [57]. In addition, reactivity control is a key step for building 3D networks with organic interlayer spacers. In the following a selection of theoretical and simulation works on adsorption of gases (mainly hydrogen) in graphene-based systems is reported. A summary of the literature is in Table 2.

4.1 Physisorption

Taking hydrogen as a paradigmatic case for molecular interactions with graphene, one can generally separate adsorption onto graphene in two main classes, chemisorption and physisorption [13]. Physisorption occurs basically barrierless via dispersive vdW interactions in case of neutral molecules. In the case of H_2, the

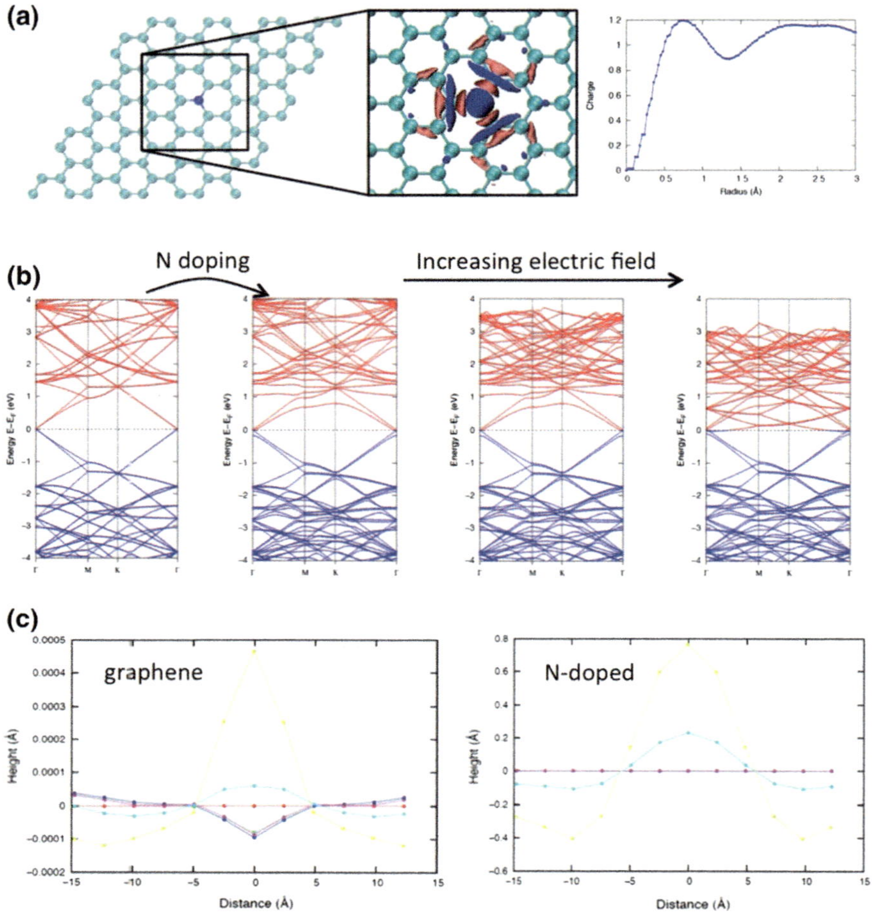

Fig. 4 Effect of N-doping on flexoelectricity of graphene (adapted from Ref. [26]). **a** $4\sqrt{3} \times 4\sqrt{3}$R30 graphene supercell with N-substitution in the center. A detail of the charge redistribution is reported (*blue* = charge accumulation, *pink* = charge depletion) and the radial distribution of electronic charge is also reported in the plot. **b** Change of the band structure as an effect of doping and of increasing electric field applied orthogonally to the sheet (max field value = 10GV/m). Besides the small band gap opening and doping level due to substitution, electric field bends the bands and enhances doping. **c** Height profile along the main diagonal of the supercell, for bare graphene (*left*) and N-doped graphene (*right*) at increasing levels of the electric field. In bare graphene the effect is very small and fluctuating also in directions. In N-doped graphene the effect is 3–4 orders of magnitude enhanced, due to electronic charge localization and symmetry breaking. A systematic study of the effects of doping on flexoelectricity is currently in the course

binding energy on bare graphene is estimated around 0.01 eV [58], bringing a precise linear dependence of the gravimetric density onto the specific surface area [14] and posing a strict upper limit to the gravimetric density for physisorption onto

Table 1 As survey of the modeling works on (morphed) graphene systems. Density Functional Theory-based calculations and simulations

System	Calculation	Main results	References
BN substituted single sheet	6 × 6 supercell (72 atoms) GGA functional	Band structure, electronic gap, and dos variation as a function of the BN relative amount and decoration shapes	[29] (2011)
Mono and Bilayers with B or N substitutions	4 × 4 supercell (32–64 atoms) PBE functional	Doping level for localized substitutions of B or N	[30] (2009)
Monolayer with BN patches	6 × 6 supercell (192 atoms) PBE functional	Band gap and doping for different size/shape of the patches and unbalance between B and N	[31] (2010)
Monolayer with dislocations and grain boundaries	Rectangular supercell of ~4 × 1.5 nm size (estimated ~150 atoms) GGA spin polarized	Electronic and transport properties dependence on the topology of defects and relative orientation of grain boundaries	[32] (2010)
Monolayer with "octite" defects isolated or in superlattices	Rectangular supercell, (144–200 atoms) DFT and GW	Band gap in different configuration is evaluated and shows marked dependence on the distribution and symmetry	[34] (2010)
Graphene on SiC (C-rich surface)	4 × 4 supercell, with buffer layer and substrate (~ 150 atoms); LSDA	Band structure, evaluation of preferential substitutions, and level of doping of different dopants (P, Al, B, N)	[35] (2011)
Graphene on SiC (C-rich surface)	2 × 2 without buffer layer PBE + empirical vdW corrections	n-type doping is naturally induced by the interaction with the substrate, but dependent on the type of passivation (with or without intercalated H), and on the inclusion of the correct vdW interactions	[36] (2011)
Graphene on SiC (Si-Rich surface)	2 × 2 with buffer layer and substrate GGA	n-type doping of 0.5 eV is observed, induced by the buffer layer–monolayer interaction	[37] (2007)
Graphene on SiC (C-rich surface)	5 × 5 with buffer layer and substrate DFT	First layer is almost metallic, with little doping	[38] (2011)
Graphene on SiC (Si-rich surface)	13 × 13 with buffer layer and substrate; LDA	The buffer is covalently bound, the first monolayer is metallic and with negligible doping	[39] (2008)

(continued)

Table 1 (continued)

System	Calculation	Main results	References
Quasi Free Standing on SiC and Li intercalated graphene	13 × 13 with complete intercaled H coverage; PBE + vdW	The QFSG is weakly interacting, with very small band gap and basically no doping	[41] (2011)
Quasi free standing graphene on SiC	13 × 13 with buffer layer and substrate and complete intercaled H coverage; PBE + vdW	The interaction between substrate and graphene is very weak, leading to uniform electronic density on rather flat sheet	[40] (2015)
Graphene on different metals (Cu, Ni, Co, Pt, Pd)	Unit and 2 × 2 cells, unstretched, with metal lattice adapted accordingly; LSDA	Graphene interaction with different metal has different character (physical or chemical), and different type and level of doping	[43] (2008)
Quasi free standing graphene on SiC with vacancy type defects	13 × 13 with buffer layer and substrate and complete intercalated H coverage; PBE + vdW	The defect strongly interacts with the substrate, producing changes in electronic and magnetic properties	[44] (2014)
Single sheet with N substitutional and pyrrolic/pyridinic defects	3 × 13 with N substitutions and/or vacancies; LDA	The effects of substitution and vacancy allow a tuning of the band gap and doping	[46] (2011)
Single sheets	Up to 400 atoms, rectangular geometry; LDA	Evaluation of flexoelectric coefficient	[53] (2008)
Single sheets fluorinated, hydrogenated, fluorinated and decorated with Li and K	Unit, 2 × 2, 3 × 3, 4 × 4 cells PBE	Evaluation of piezoelectric coefficient in functionalized graphene	[54] (2012)
Single sheet, with saturated vacancy defects	Rectangular (10 × 5)	Evaluation of piezoelectric coefficient in graphene with vacancies	[56] (2012)

graphene at 1–2% at room temperature. Besides the obvious strategy of working at cryogenic temperature, in several theoretical works physical interactions were shown enhanced within nanocavities produced by rippling [58], within multilayers with specific (nanometric) interlayer spacing [59, 60] or with nanosized perforations of the sheet [61].

Table 2 A survey of literature on chemi and physisorption of H_2 onto and in graphene

System	Calculation	Main results	Ref
Single layer and multilayer, rippled	Rectangular cell, 180 atoms; PBE	H chemisorbs on convexities and is unstable within concavities. H_2 physisorbs preferentially within concavities	[58] (2011)
Multilayer with different layer spacing, with molecular hydrogen	30 × 30 supercell; Accurate H2-graphene potential fitted on post-Hartree–Fock theories; Quantum treatment of H_2 dynamics	For specific interlayer spacing, the hydrogen adsorption is enhanced	[59] (2005)
Multilayer graphene-oxide framework pillared with diboronic-derived molecules exposed to CO_2	~2 nm × 2 nm × 2 nm DFT-PBE, DFTB Classical MMFF94	Using realistic framework models with the correct spacing and corrugation enhances selectivity and the adsorption coefficient	[60] (2015)
Multilayer perforated graphene exposed to H2	Nanosized supercells. Classical LJ interactions for H2-gr interaction Grand Canonical Monte Carlo dynamics for hydrogen	Graphene perforation effectively enhances the active surface and increase the physisorption capability	[61] (2015)
Sheets with H, O, OH, CH_3 ad-groups	30 × 30 supercell; GGA	Evaluation of binding energies and hopping barriers	[66] (2009)
Single sheet exposed to H2 embedded in electric fields	2 × 2 supercell; LDA	Barriers are lowered by EF orthogonal to the sheet	[71] (2010)
Single sheet exposed to H_2, N-doped, embedded in EF	4 × 4 supercell; RPBE	The effect of N-doping and EF cooperate, leading to substantial decrease the chemisorption barrier	[73] (2012)
Sheets corrugated with sinusoidal out-of-plane deformations of different wavelength	Rectangular supercells 28–84 atoms; PBE	The fluorination and hydrogenation energy depends on local curvature	[77] (2011)

From the methodological point of view, it is interesting to observe that these works address the problem of physisorption combining different methodologies besides DFT (see Table 2), including classical approaches based on empirical Force

Fields, different types of dynamical treatment of hydrogen (QM dynamics, Monte Carlo and thermodynamic evaluation of the adsorption). This is basically due to two circumstances: first, gravimetric density evaluation must be treated at the statistical level, which requires large system and sampling methods proper to this aim. Second, the dispersive interactions are very elusive and difficult to treat with standard DFT methods. Commonly used LDA and GGA Density Functionals are based on the (semi)local representation of exchange-correlation, insufficient to address instantaneous fluctuations of the electron density. As a consequence, these functionals fail in reproducing the medium-long range behavior of interatomic interactions [62, 63], which was corrected with a number of semiempirical schemes. In addition, hydrogen, the lightest element, calls a quantum treatment of its dynamics. Because all of these circumstances make the simulation unaffordable on large systems, recursion to combination of high accuracy/high cost with lower accuracy/low cost methods in multi-scale combined approach is compulsory. Overall, however, the lesson one learns from this analysis is that cavities, convexities, porosity or in general structures enhance the physisorption at best when they are at the nanoscale (specifically ~1 nm). Therefore nanoscale structuring must be considered a guideline also in designing graphene-based 3D networks.

4.2 Manipulation of Reactivity

In contrast to physisorption, chemisorption on graphene is a barrier-driven process, requiring a change in hybridization from sp^2 to sp^3 [64, 65]. The barrier for atomic H is estimated around 0.3 eV, but the barrier for H_2, involving dissociative chemisorption, is ~1.5 eV per atom. Once adsorbed H can hop from site to site with a barrier of ~0.75 eV [66]. These processes display slow kinetic at room temperature.

Enhancement of reactivity can be obtained by perturbing the π delocalization in several ways, which mostly superimpose to those already mentioned to manipulate the electronic structure. Structural and substitutional [67–70] defects generally constitute hot spots of reactivity. The action mechanism is usually ascribed to the presence of localized states or doping due to the defect, which are able to induce dissociation of the molecule and subsequent adsorption of the reactive radicals, resulting in a reduction of the dissociative chemisorption barrier. External electric fields were shown to act as catalysts for this process [71, 72], being able to deform the already perturbed orbitals. This effect is synergetic with that of substitutional doping [73–75] promising fast kinetics, large GD (6.73%) at room temperature and the possibility of using electric field for switch uptake/release of molecules.

The reactivity enhancement effect of purely structural defects is partially due to the local distortion from flatness, which induces a protruding and partial pyramidalization of a C site, favoring the sp^2 to sp^3 transition. In fact, significant lowering of the barriers for H_2 dissociation were previously observed on nanotubes [76], while stabilization of the adsorbate was observed on corrugated graphene without defects [58, 77, 78]. This indicates that control of local curvature can induce control of graphene decoration, upload and even release, as demonstrated by a simulation showing H_2 release by curvature inversion [58]. A side (but related) effect, is that the band gap also depends on mechanical deformation, therefore rippling [28] or stretching [79] could be used directly to manipulate electronic properties, and indirectly by curvature induced chemical functionalization.

Because chemisorption involves much larger energies than physisorption, in a first approximation vdW corrections to standard DFT schemes play a less important role. However, they become crucial in the approaching phase of the molecule, when different orientations and locations with respect to C atoms determine the effective reaction path and barrier for the chemisorption reaction [80–82].

4.3 Metal-Mediated Adsorption

Metal mediated adsorption of hydrogen is often located [83] in between chemisorption and physisorption. This interaction is specifically mediated by d orbitals in transition (e.g., Ti [84]) or heavy metals (e.g. Pd [85, 86] which also works as catalyst and dissociates the molecule producing a "spillover" effect of atomic H onto graphene), while it is often described by an "enhanced" vdW interaction for lighter metals such as Li and Ca [87]. In all cases the interaction energy is increased of at least one order of magnitude, which suggests useful applications in storage. However the problem is turned in how to control the distribution of metal onto graphene. In fact in most cases metals tend to form clusters, reducing the active surface for adsorption with respect to total mass.

Again, combination of different kinds of graphene manipulation has a key role. For instance, it was shown that inducing N-defects onto graphene before exposure to Ti can reduce the cluster size, optimizing the active surface for adsorption [88]. Combination of Li decoration, N-doping and electric field predict reversibility of hydrogenation in a DFT study [89].

Morphology manipulations have a diverse effect on the selectivity of the adsorption process of molecules on graphene, so that it can be used for gas sensor applications. Doping graphene with heteroatoms can result in a different amount of valence electrons, and usually protrusion (e.g., with Al, Si, P, S, Mn, Cr) [90, 91] from the graphene surface and creation of reactive binding sites for molecular adsorption. In general, as in the case of physisorption and more than for chemisorption, the vdW interactions have a main role, therefore the use of appropriate functional for their representation is crucial [92].

5 Dynamical Morphing

Reversibility of graphene manipulation is a key property in tasks such as hydrogen storage. Mechanical deformations are the most reversible. In fact the reversible storage using the reactivity dependence on local curvature was demonstrated in simulations [58]. Therefore, the problem is turned on how to control local curvature of the graphene sheet. One possibility is to exploit the natural curvature of supported graphene: on given substrates, such as SiC, graphene shows multi-stable patterns of curvature [93], allowing the possibility of switching between each other by changing external environmental conditions (e.g., temperature, electric fields).

However, graphene offers the unique possibility of changing the local curvature dynamically, exploiting the naturally occurring out-of-plane wave-like deformations, i.e., the flexural phonons (or acoustic z polarized branches, i.e., ZA phonons). The capability of those vibrations to detach chemisorbed hydrogen and transport and pump it though a multilayer system was demonstrated by DFT [58] and classical MD simulations based on empirical FFs [11]. Clearly, for such applications, the key issue is how to create and maintain coherent ZA phonons of specific wavelength and amplitude.

Many efforts were devoted to the calculation the phonon dispersion curves of graphene. The reason is that due to the 2d nature of this material, the vibrational properties play a crucial role on its thermal and mechanical properties. Much attention has been reserved to the ZA mode, particularly important in 2d materials as graphene [94]. Its dispersion law is parabolic, implying that these modes are the lowest energies of the spectrum, and are the easiest to be excited. The very low phonon–phonon scattering rate and the large thermal population have as a consequence that the flexural phonons give a fundamental contribution to thermal conductivity both for graphene monolayer [95] and multilayers [96]. Moreover, they are responsible of the negative thermal expansion coefficient observed in a wide temperature interval, up to an inversion temperature value, which is still controversial [94–96].

Among the most used computational approaches we can mention the elastic continuum model [97, 98], the first principles DFT also with Perdew–Burke–Ernzerhof generalized gradient approximation [99–101], and the molecular dynamics simulations associated with the use of empirical interatomic potentials for the interactions between carbon atoms, which allows to treat larger systems [102]. The most effective have been demonstrated the Long-range Carbon Bond-Order Potential (LCBOPII) [103] and the parameterization by Lindsay et al. [104]. of the Tersoff potential [105], which has been also used to calculate the thermal conductivity of graphene [106, 107]. Empirical interatomic potentials have been widely used also for the investigation of the mechanical and elastic properties of graphene sheets, in particular for the elastic bending modulus [108], the thickness and Young's modulus [109, 110], also on Si-terminated SiC [111] and 6H-SiC(0001) [112]. A survey of simulation studies about dinamical properties and mechanical manipulation of graphene is reported in Table 3.

Table 3 Vibrational mechanical and thermal properties of graphene systems, theoretical approaches

System	Calculation	Main results	Ref
Suspended graphene sheet	0.4 × 20 nm supercell AIREBO FF	Suspended graphene deflects in the presence of an electric field orthogonal to the sheet	[53] (2010)
Graphene single layer	MD with classical force field	Dynamics of flexural modes and its relationship with thermal and mechanical properties	[94] (2015)
Graphene single layer	MD with classical force field	Lattice thermal conductivity of graphene is dominated by contributions from the flexural phonon mode	[95] (2010)
Graphene multilayer and graphite	MD with classical force field	Lattice thermal conductivity of graphene is dominated by ZA mode with a lower thermal conductivity due to a break of a selection rule on phonon–phonon scattering	[96] (2011)
Graphene nanoribbon	2d continuum model	Acoustic phonon dispersion curves in the case of fixed and free boundaries	[97] (2011)
Graphene nanoribbon	Elastic continuum model	Optical phonon dispersion curves	[98] (2009)
Graphene single layer	DFT-LDA, plane waves basis set. rectangular supercell of 128 atoms	Phonon dispersion relations	[99] (2003)
Graphene single layer	DFT and DFPT, PBE	Structural, dynamical, and thermodynamic properties, phonon dispersion curves	[100] (2005)
Graphene monolayer, bilayer, trilayer	DFPT	Phonon dispersion curves; optical phonon $E2\ g$ mode at Γ splits into two and three doubly degenerate branches for bilayer and trilayer graphene	[101] (2008)
Graphene monolayer	MD with empirical potentials (Tersoff, Tersoff-Lindsay, LCBOP, AIREBO). 20 × 20 unit cells	Phonon dispersion curves and modification with temperature. Tersoff potential with modified parameterization shows the most physically sound behavior	[102] (2015)
Graphene monolayer and graphite	Standard lattice dynamics with LCBOPII empiric interatomic potential	Phonon dispersion and bending rigidity	[103] (2011)

(continued)

Table 3 (continued)

System	Calculation	Main results	Ref
Graphene monolayer	Tersoff and Brenner potential	Optimization of and modification of FF for the accurate reproduction of phonons	[104] (2010)
Graphene monolayer	Tersoff potential, parameters in [11]	Calculation of thermal conductivity	[106] (2012)
Graphene monolayer	Monte Carlo with empiric interatomic potentials. isothermal–isobaric ensemble, periodic boundary conditions along x, y, reparameterized long-range carbon bond-order potential LBOP and LCBOP	Thermodynamic properties; empirical potentials limited to nearest-neighbor interactions give rather dispersed results	[107] (2014)
Graphene monolayer	Brenner (first generation)	Analytic form for the bending modulus	[108] (2009)
Graphene monolayer and single-wall nanotube	Analytic from interatomic potential	tension and bending rigidity directly from the interatomic potential	[109] (2006)
Graphene monolayer	Continuum approach; Molecular mechanics, REBO interatomic empirical potential	Nonlinear elastic properties under uniaxial stretch and tension	[110] (2012)
Graphene layer on Si terminated SIC	environment-dependent interatomic empirical potential (EDIP) Graphene layer basic cell of 338 atoms	Determination of the atomistic structure of the graphene buffer layer on Si-terminated SiC. The solution of minimal energy forms a hexagonal pattern composed of stuck regions separated by unbonded rods that release the misfit with the SiC surface	[111] (2010)
Graphene layer	Molecular dynamics simulation, Tersoff empirical interatomic potentials for C–C, Si–Si, and Si–C interactions	Two layers of carbon atoms of the 6H-SiC(0001) subsurface after sublimation of Si atoms undergo a transformation from a diamondlike phase to a graphene-like structure at annealing temperature above 1500 K	[112] (2008)

6 Conclusions and Perspectives

Morphology manipulation of graphene requires flexible experimental and modeling approaches. In particular on the modeling side, both accuracy in the representation of the phenomena, and the use of model systems very large compared to the atomic scale are required. The former calls for the use of at least DFT based schemes, most often corrected for long-range electron correlation effects to better account for specific electronic properties or for the dispersive part of vdW interaction; in specific cases the recursion to quantum mechanical treatment of nuclei is also required. On the other hand, large model systems are needed to analyze systems at the nano-micro scale with defects and functionalization, which have a low level of symmetry. The evaluation of thermodynamic properties is also often needed, requiring extensively long simulations. This is often incompatible with the high computational cost of accurate methods, therefore empirical treatment of interactions is used. In some cases the two methods, ab initio and empirical, are mixed at several levels, intrinsically (such as in the empirical vdW corrections to DFT) or explicitly.

Ab initio level currently allows addressing the 10^5 atoms scale (corresponding approximately to the 100 nm size in graphene) in short simulations. Considering the exponential increase of computer power ensured by Moore's law, the μm scale can be reached in less than 10 years, while it is already feasible, in principle, using empirical approaches on computing systems with extensive parallelization.

However, as the scales feasible in simulations increase, new questions emerge on the reliability of the theories underlying the simulation. This is especially true for more empirical approaches, because the parameterization of interactions are usually tested on smaller/shorter size/time scales, but also for more ab initio approaches such as DFT, which always include some hidden level of empiricism. Other sources of "systematic" errors in modeling might arise in the algorithms used for sampling of the conformations and phase space of the system and in the large scale average or thermodynamic properties.

In conclusion, as the size of the model system increases, the comparison with measurements becomes more and more important. The sizes of experiment and of simulations tend to meet – the first coming from bottom, the second from top—at the meso-scale. This allows a direct comparison of measured observable with the corresponding evaluated quantity. Clearly this brings an advantage in the interpretation of experiments, but also a feedback on the theory, allowing creating models more adherent to reality.

Acknowledgements We gratefully acknowledge financial support by funding from the European Union's Horizon 2020: the Marie Sklodowska-Curie grant agreement No 657070 and the graphene Core 1, Grant Agreement No. 696656; the CINECA award ISCRAC "ElMaGRe"—HP10C6U75 M, 2015 and ISCRAC "HBG", 2013 and PRACE "Tier0" award Pra07_1544 for resources on FERMI (IBM Blue Gene/Q@CINECA, Bologna Italy), and the CINECA staff for technical support. We gratefully thank Dr Vittorio Pellegrini and Prof P. Giannozzi for useful discussions.

References

1. Geim, A.K., Novoselov, K.S.: The rise of graphene. Nat. Mater. **6**, 183–191 (2007)
2. Castro Neto, A.H., Guinea, F., Peres, N.M.R., Novoselov, K.S., Geim, A.K.: The electronic properties of graphene. Rev. Mod. Phys. **81**, 109 (2009)
3. Novoselov, K.S., Geim, A.K., Morozov, S.V., Jiang, D., Katsnelson, M.I., Grigorieva, I.V., Dubonos, S.V., Firsov, A.A.: Two-dimensional gas of massless Dirac fermions in graphene. Nature **438**, 197–200 (2005)
4. Lee, C., Wei, X., Kysar, J.W., Hone, J.: Measurement of the elastic properties and intrinsic strength of monolayer graphene. Science **321**(5887), 385 (2008)
5. Nicklow, R., Wakabayashi, N., Smith, H.G.: Lattice dynamics of pyrolytic graphite. Phys. Rev. B **5**(12), 4951 (1972)
6. Kudin, K.N., Scuseria, G.E., Yakobson, B.I.: C2F, BN and C nanoshell elasticity from ab initio computations. Phys. Rev. B **64**, 235406 (2001)
7. Fasolino, A., Los, J.H., Katsnelson, M.I.: Intrinsic ripples in graphene. Nat. Mater. **6**, 858–861 (2007)
8. Jomehzadeh, E., Afshar, M.K., Galiotis, C., Shi, X., Pugno, N.M.: Nonlinear softening and hardening nonlocal bending stiffness of an initially curved monolayer graphene. Int. J. Non Linear Mech. **56**, 123–131 (2013)
9. Fiori, G., Bonaccorso, F., Iannaccone, G., Palacios, T., Neumaier, D., Seabaugh, A., Banerjee, S.K., Colombo, L.: Electronics based on two-dimensional materials. Nat. Nanotechnol. **9**, 768–779 (2014)
10. Siochi, E.J.: Graphene in the sky and beyond. Nat. Nanotechnol. **9**, 745–747 (2014)
11. Camiola, V.D., Farchioni, R., Pellegrini, V., Tozzini, V.: Hydrogen transport within graphene multilayers by means of flexural phonons. 2D Mater. **2**(1), 014009 (2015)
12. De Andres, P.L., Guinea, F., Katsnelson, M.I.: Bending modes, anharmonic effects, and thermal expansion coefficient in single-layer and multilayer graphene. Phys. Rev. B **86**, 144103 (2012)
13. Tozzini, V., Pellegrini, V.: Prospects for hydrogen storage in graphene. PCCP **15**(1), 80–89 (2013)
14. Klechikov, A., Mercier, G., Sharifi, T., Baburin, I.A., Seifert, G., Talyzin, A.V.: Hydrogen storage in high surface area graphene scaffolds. Chem. Commun. **51**, 15280–15283 (2015)
15. Wang, H., Yuan, X., Zeng, G., Wu, Y., Liu, Y., Jiang, Q., Gu, S.: Three dimensional graphene based materials: Synthesis and applications from energy storage and conversion to electrochemical sensor and environmental remediation. Adv. Colloid Interface Sci. **221**, 41–59 (2015)
16. Mercier, G., Klechikov, A., Hedenström, M., Johnels, D., Baburin, I.A., Seifert, G., Mysyk, R., Talyzin, A.V.: Porous graphene oxide/diboronic acid materials: structure and hydrogen sorption. J. Phys. Chem. C **119**(49), 27179–27191 (2015)
17. Xia, Z., Leonardi, F., Gobbi, M., Liu, Y., Bellani, V., Liscio, A., Kovtun, A., Li, R., Feng, X., Orgiu, E., Samori, P., Treossi, E., Palermo, V.: Electrochemical Functionalization of graphene at the nanoscale with self-assembling diazonium salts. ACS Nano, Just Accepted Manuscript (2016). doi:10.1021/acsnano.6b03278
18. Jiang, H., Lee, P.S., Li, C.: 3D carbon based nanostructures for advanced supercapacitors. Energy Environ. Sci. **6**, 41 (2013)
19. Georgakilas, V., Spinato, C., Ménard-Moyon, C., Bianco, A.: Chemical functionalization of graphene for biomedical applications. In: Functionalization of Graphene. Wiley on-line library (2014). doi:10.1002/9783527672790.ch4
20. Tozzini, V.: Multiscale modeling of proteins. Acc. Chem. Res. **43**(2), 220–230 (2010)
21. Camiola, V.D., Farchioni, R., Cavallucci, T., Rossi, A., Pellegrini, V., Tozzini, V.: Hydrogen storage in rippled graphene: perspectives from multi-scale simulations. Front. Mater. **2**, 00003 (2015)

22. Akshay, S., Srikanth, B., Kumar, A., Dasmahapatra, A.K.: Coarse-grain molecular dynamics study of fullerene transport across a cell membrane. J. Chem. Phys. **143**(2) (2015)
23. Trovato F., Tozzini, V.: Diffusion within the cytoplasm: a mesoscale model of interacting macromolecules. Biophys. J. **107**(11), 2579–2591 (2015)
24. Jomehzadeh, E., Pugno, N.M.: Bending stiffening of graphene and other 2D materials via controlled rippling. Compos. Part B **83**, 194–202 (2015)
25. Konschuh, S., Gmitra, M., Fabian, J.: Tight-binding theory of the spin-orbit coupling in graphene. Phys. Rev. B **82**, 245412 (2010)
26. Cavallucci, T.: Density Functional Theory simulations of the electromechanical properties of naturally corrugated epitaxial graphene. Master Thesis in Physics University of Pisa (2014)
27. Trevisanutto, P.E., Giorgetti, C., Reining, L., Ladisa, M., Olevano, V.: Ab initio GW many-body effects in graphene. Phys. Rev. Lett. **101**, 226405 (2008)
28. Rossi, A.: Corrugated graphene hydrogenation with Density Functional Theory based simulations. Master Thesis in Physics University of Pisa (2014)
29. Shinde, P.P., Kumar, V.: Direct band gap opening in graphene by BN doping: ab initio calculations. Phys. Rev. B **84**, 125401 (2011)
30. Panchakarla, L.S., Subrahmanyam, K.S., Saha, S.K., Govindaraj, A., Krishnamurthy, H.R., Waghmare, U.V., Rao, C.N.R.: Synthesis, structure, and properties of boron- and nitrogen-doped graphene. Adv. Mater. **21**, 4726–4730 (2009)
31. Xu, B., Lu, Y.H., Feng, Y.P., Lin, J.Y.: Density functional theory study of BN-doped graphene superlattice: role of geometrical shape and size. J. Appl. Phys. **108**, 073711 (2010)
32. Yazyev, O.V., Louie, S.G.: Topological defects in graphene: dislocations and grain boundaries. Phys. Rev. B **81**, 195420 (2010)
33. Yazyev, O.V., Louie, S.G.: Electronic transport in polycrystalline graphene. Nat. Mater. **9** (10), 806–809 (2010)
34. Appelhans, D.J., Carr, L.D., Lusk, M.T.: Embedded ribbons of graphene allotropes: an extended defect perspective. New J. Phys. **12**(12), 125006 (2010)
35. Huang, B., Xiang, H.J., Wei, S.H.: Controlling doping in graphene through a SiC substrate: a first-principles study. Phys. Rev. B **83**(16), 161405 (2011)
36. Jayasekera, T., Xu, S., Kim, K.W., Nardelli, M.B.: Electronic properties of the graphene/6H-SiC ($000\bar{1}$) interface: a first-principles study. Phys. Rev. B **84**(3), 035442 (2011)
37. Varchon, F., Feng, R., Hass, J., Li, X., Nguyen, B.N., Naud, C., Mallet, P., Veuillen, J.-Y., Berger, C., Conrad, E.H., Magaud, L.: Electronic structure of epitaxial graphene layers on SiC: effect of the substrate. Phys. Rev. Lett. **99**(12), 126805 (2007)
38. Deretzis, I., La Magna, A.: Single-layer metallicity and interface magnetism of epitaxial graphene on SiC ($000\bar{1}$). Appl. Phys. Lett. **98**(2), 3113 (2011)
39. Kim, S., Ihm, J., Choi, H.J., Son, Y.W.: Origin of anomalous electronic structures of epitaxial graphene on silicon carbide. Phys. Rev. Lett. **100**(17), 176802 (2008)
40. Sforzini, J., Nemec, L., Denig, T., Stadtmüller, B., Lee, T.L., Kumpf, C., Soubatch, S., Starke, U., Rinke, P., Blum, V., Bocquet, F.C., Tautz, F.S.: Approaching truly freestanding graphene: the structure of hydrogen-intercalated graphene on 6 H−SiC (0001). Phys. Rev. Lett. **114**(10), 106804 (2015)
41. Deretzis, I., La Magna, A.: Role of covalent and metallic intercalation on the electronic properties of epitaxial graphene on SiC (0001). Phys. Rev. B **84**(23), 235426 (2011)
42. Forti, S., Emtsev, K.V., Coletti, C., Zakharov, A.A., Riedl, C., Starke, U.: Large-area homogeneous quasifree standing epitaxial graphene on SiC (0001): electronic and structural characterization. Phys. Rev. B **84**(12), 125449 (2011)
43. Giovannetti, G., Khomyakov, P.A., Brocks, G., Karpan, V.M., Van den Brink, J., Kelly, P. J.: Doping graphene with metal contacts. Phys. Rev. Lett. **101**(2), 026803 (2008)
44. Sławińska, J., Cerda, J.I.: The role of defects in graphene on the H-terminated SiC surface: not quasi-free-standing any more. Carbon **74**, 146–152 (2014)

45. Jeong, H.M., Lee, J.W., Shin, W.H., Choi, Y.J., Shin, H.J., Kang, J.K., Choi, J.W.: Nitrogen-doped graphene for high-performance ultracapacitors and the importance of nitrogen-doped sites at basal planes. Nano Lett. **11**(6), 2472–2477 (2011)
46. Fujimoto, Y., Saito, S.: Formation, stabilities, and electronic properties of nitrogen defects in graphene. Phys. Rev. B **84**(24), 245446 (2011)
47. Sofo, J.C., Chaudhari, A.S., Barber, G.D.: Graphane: a two-dimensional hydrocarbon. Phys. Rev. B **75**(15), 153401 (2007)
48. Tozzini, V., Pellegrini, V.: Electronic structure and Peierls instability in graphene nanoribbons sculpted in graphane. Phys. Rev. B **81**(11), 113404 (2010)
49. Goler, S., Coletti, C., Tozzini, V., Piazza, V., Mashoff, T., Beltram, F., Pellegrini, V., Heun, S.: The influence of graphene curvature on hydrogen adsorption: towards hydrogen storage devices. J. Phys. Chem. C **117**(22), 11506–11513
50. Boukhvalov, D.W., Katsnelson, M.I., Lichtenstein, A.I.: Hydrogen on graphene: electronic structure, total energy, structural distortions and magnetism from first-principles calculations. Phys. Rev. B **77**(3), 035427 (2008)
51. Rossi, A., Piccinin, S., Pellegrini, V., de Gironcoli, S., Tozzini, V.: Nano-scale corrugations in graphene: a density functional theory study of structure, electronic properties and hydrogenation. J. Phys. Chem. C **119**(14), 7900–7910 (2015)
52. Haberer, D., Vyalikh, D.V., Taioli, S., Dora, B., Farjam, M., Fink, J., Marchenko, D., Pichler, T., Ziegler, K., Simonucci, S., Dresselhaus, M.S., Knupfer, M., Büchner, B., Grüneis, A.: Tunable band gap in hydrogenated quasi-free-standing graphene. Nano Lett. **10**(9), 3360–3366 (2010)
53. Wang, Z., Philippe, L., Elias, J.: Deflection of suspended graphene by a transverse electric field. Phys. Rev. B **81**(15), 155405 (2010)
54. Kalinin, S.V., Meunier, V.: Electronic flexoelectricity in low-dimensional systems. Phys. Rev. B **77**(3), 033403 (2008)
55. Ong, M.T., Reed, E.J.: Engineered piezoelectricity in graphene. ACS Nano **6**(2), 1387–1394 (2012)
56. Chandratre, S., Sharma, P.: Coaxing graphene to be piezoelectric. Appl. Phys. Lett. **100**(2), 023114 (2012)
57. Quesnel, E., Roux, F., Emieux, F., Faucherand, P., Kymakis, E., Volonakis, G., Giustino, F., Martín-García, B., Moreels, I., Gürsel, S.A., Yurtcan, A.B., Di Noto, V., Talyzin, A., Baburin, I., Tranca, D., Seifert, G., Crema, L., Speranza, G., Tozzini, V., Bondavalli, P., Pognon, G., Botas, C., Carriazo, D., Singh, G., Rojo, T., Kim, G., Yu, W., Grey C.P., Pellegrini, V.: Graphene-based technologies for energy applications, challenges and perspectives. 2D Mater. **2**(3), 030204 (2015)
58. Tozzini, V., Pellegrini, V.: Reversible hydrogen storage by controlled buckling of graphene layers. J. Phys. Chem. C **115**(51), 25523–25528 (2011)
59. Patchkovskii, S., John, S.T., Yurchenko, S.N., Zhechkov, L., Heine, T., Seifert, G.: Graphene nanostructures as tunable storage media for molecular hydrogen. Proc. Natl. Acad. Sci. U.S.A. **102**(30), 10439–10444 (2005)
60. Garberoglio, G., Pugno, N.M., Taioli, S.: Gas adsorption and separation in realistic and idealized frameworks of organic pillared graphene: a comparative study. J. Phys. Chem. C **119**(4) 1980–1987 (2015)
61. Baburin, I.A., Klechikov, A., Mercier, G., Talyzin, A., Seifert, G.: Hydrogen adsorption by perforated graphene. Int. J. Hydrogen Energy **40**(20), 6594–6599 (2015)
62. Pérez-Jordá, J., Becke, A.D.: A density-functional study of van der Waals forces: rare gas diatomics. Chem. Phys. Lett. **233**(1), 134–137 (1995)
63. Kristyán, S., Pulay, P.: Can (semi) local density functional theory account for the London dispersion forces? Chem. Phys. Lett. **229**(3), 175–180 (1994)
64. Miura, Y., Kasai, H., Diño, W., Nakanishi, H., Sugimoto, T.: First principles studies for the dissociative adsorption of H_2 on graphene. J. Appl. Phys. **93**, 3395–3400 (2003)
65. Dzhurakhalov, A.A., Peeters, F.M.: Structure and energetics of hydrogen chemisorbed on a single graphene layer to produce graphane. Carbon **49**(10), 3258–3266 (2011)

66. Wehling, T.O., Katsnelson, M.I., Lichtenstein, A.I.: Impurities on graphene: midgap states and migration barriers. Phys. Rev. B **80**(8), 085428 (2009)
67. Zhao, L., He, R., Rim, K.T., Schiros, T., Kim, K.S., Zhou, H., Gutiérrez, C., Chockalingam, S.P., Arguello, C.J., Pálová, L., Nordlund, D., Hybertsen, M.S., Reichman, D.R., Heinz, T. F., Kim, P., Pinczuk, A., Flynn, G.W., Pasupathy, A.N.: Visualizing individual nitrogen dopants in monolayer graphene. Science **333**(6045), 999–1003 (2011)
68. Kim, B.H., Hong, S.J., Baek, S.J., Jeong, Y.H., Park, N., Lee, M., Lee, S.W., Park, M., Chu, S.W., Shin, H.S., Lim, J., Lee, J.C., Jun, Y., Park, Y.W.: N-type graphene induced by dissociative H_2 adsorption at room temperature. Sci. Rep. **2**, 690 (2012)
69. Gallouze, M., Kellou, A., Drir, M.: Adsorption isotherms of H_2 on defected graphene: DFT and Monte Carlo studies. Int. J. Hydrogen Energy **41**, 5522–5530 (2016)
70. Seydou, M., Lassoued, K., Tielens, F., Maurel, F., Raouafi, F., Diawara, B.: A DFT-D study of hydrogen adsorption on functionalized graphene. RSC Adv. **5**, 14400–14406 (2015)
71. Ao, Z.M., Peeters, F.M.: Electric field: a catalyst for hydrogenation of graphene. Appl. Phys. Lett. **96**(25), 253106 (2010)
72. Shi, S., Hwang, J.-Y., Li, X., Sun, X., Lee, B.I.: Enhanced hydrogen sorption on carbonaceous sorbents under electric field. Int. J. Hydrogen Energy **35**(2), 629–631 (2010)
73. Ao, Z.M., Hernández-Nieves, A.D., Peeters, F.M., Li, S.: The electric field as a novel switch for uptake/release of hydrogen for storage in nitrogen doped graphene. Phys. Chem. Chem. Phys. **14**(4), 1463–1467 (2012)
74. Ao, Z., Li, S.: Hydrogenation of graphene and hydrogen diffusion behavior on graphene/graphane interface. In: Gong, J.R. (ed.) Graphene Simulation. Intech; Chapt. 4:53–74 (2011)
75. Ao, Z., Li, S.: Electric field manipulated reversible hydrogen storage in graphene studied by DFT calculations. Phys. Status Solidi A **211**(2), 351–356 (2014)
76. Costanzo, F., Silvestrelli, P.L., Ancilotto, F.: Physisorption, diffusion, and chemisorption pathways of H_2 molecule on graphene and on (2,2) carbon nanotube by first principles calculations. J. Chem. Theory Comput. **8**, 1288–1294 (2012)
77. Gao, X., Wang, Y., Liu, X., Chan, T.L., Irle, S., Zhao, Y., Zhang, S.B.: Regioselectivity control of graphene functionalization by ripples. Phys. Chem. Chem. Phys. **13**(43), 19449–19453 (2011)
78. Boukhvalov, D.W., Katsnelson, M.I.: Enhancement of chemical activity in corrugated graphene. J. Phys. Chem. C **113**, 14176–14178 (2009)
79. McKay, H., Wales, D.J., Jenkins, S.J., Verges, J.A., de Andres, P.L.: Hydrogen on graphene under stress: molecular dissociation and gap opening. Phys. Rev. B **81**, 075425 (2010)
80. Boukhvalov, D.W., Katsnelson, M.I., Lichtenstein, A.I.: Hydrogen on graphene: Electronic structure, total energy, structural distortions and magnetism from first-principles calculations. Phys. Rev. B **77**(3), 035427 (2008)
81. Leenaerts, O., Partoens, B., Peeters, F.M.: Adsorption of H_2O, NH_3, CO, NO_2, and NO on graphene: a first-principles study. Phys. Rev. B **77**(12), 125416 (2008)
82. Wehling, T.O., Katsnelson, M.I., Lichtenstein, A.I.: Adsorbates on graphene: impurity states and electron scattering. Chem. Phys. Lett. **476**(4), 125–134 (2009)
83. Chan, K.T., Neaton, J.B., Cohen, M.L.: First-principles study of metal adatom adsorption on graphene. Phys. Rev. B **77**(23), 235430 (2008)
84. Takahashi, K., Isobe, S., Omori, K., Mashoff, T., Convertino, D., Miseikis, V., Coletti, C., Tozzini, V., Heun, S.: Revealing the multibonding state between hydrogen and graphene-supported Ti clusters. J. Phys. Chem. C **120**, 12974–12979 (2016)
85. Parambhath, V.B., Nagar, R., Sethupathi, K., Ramaprabhu, S.: Investigation of spillover mechanism in palladium decorated hydrogen exfoliated functionalized graphene. J. Phys. Chem. C **115**(31), 15679–15685 (2011)
86. Gierz, I., Riedl, C., Starke, U., Ast, C.R., Kern, K.: Atomic hole doping of graphene. Nano Lett. **8**(12), 4603–4607 (2008)
87. Dai, J., Yuan, J., Giannozzi, P.: Gas adsorption on graphene doped with B, N, Al, and S: a theoretical study. Appl. Phys. Lett. **95**(23), 232105 (2009)

88. Mashoff, T., Convertino, D., Miseikis, V., Coletti, C., Piazza, V., Tozzini, V., Beltram, F., Heun, S.: Increasing the active surface of titanium islands on graphene by nitrogen sputtering. Appl. Phys. Lett. **106**(8), 083901 (2015)
89. Lee, S., Lee, M., Chung, Y.-C.: Enhanced hydrogen storage properties under external electric fields of N-doped graphene with Li decoration. Phys. Chem. Chem. Phys. **15**, 3243 (2013)
90. Dai, J., Yuan, J.: Adsorption of molecular oxygen on doped graphene: atomic, electronic, and magnetic properties. Phys. Rev. B **81**(16), 165414 (2010)
91. Zhang, Y.H., Chen, Y.B., Zhou, K.G., Liu, C.H., Zeng, J., Zhang, H.L., Peng, Y.: Improving gas sensing properties of graphene by introducing dopants and defects: a first-principles study. Nanotechnology **20**(18), 185504 (2009)
92. Wong, J., Yadav, S., Tam, J., Singh, C.V.: A van der Waals density functional theory comparison of metal decorated graphene systems for hydrogen adsorption. J. Appl. Phys. **115**, 224301 (2014)
93. Cavallucci, T., Tozzini, V.: Multistable rippling of graphene on SiC: a density functional theory study. J. Phys. Chem. C **120**, 7670–7677 (2016)
94. Jiang, J.W., Wang, B.S., Wang, J.S., Park, H.S.: A review on the flexural mode of graphene: lattice dynamics, thermal conduction, thermal expansion, elasticity and nanomechanical resonance. J. Phys.: Condens. Matter **27**(8), 083001 (2015)
95. Lindsay, L., Broido, D.A., Mingo, N.: Flexural phonons and thermal transport in graphene. Phys. Rev. B **82**(11), 115427 (2010)
96. Lindsay, L., Broido, D.A., Mingo, N.: Flexural phonons and thermal transport in multilayer graphene and graphite. Phys. Rev. B **83**(23), 235428 (2011)
97. Droth, M., Burkard, G.: Acoustic phonons and spin relaxation in graphene nanoribbons. Phys. Rev. B **84**(15), 155404 (2011)
98. Qian, J., Allen, M.J., Yang, Y., Dutta, M., Stroscio, M.A.: Quantized long-wavelength optical phonon modes in graphene nanoribbon in the elastic continuum model. Superlattices Microstruct. **46**(6), 881–888 (2009)
99. Dubay, O., Kresse, G.: Accurate density functional calculations for the phonon dispersion relations of graphite layer and carbon nanotubes. Phys. Rev. B **67**(3), 035401 (2003)
100. Mounet, N., Marzari, N.: First-principles determination of the structural, vibrational and thermodynamic properties of diamond, graphite, and derivatives. Phys. Rev. B **71**(20), 205214 (2005)
101. Yan, J.A., Ruan, W.Y., Chou, M.Y.: Phonon dispersions and vibrational properties of monolayer, bilayer, and trilayer graphene: density-functional perturbation theory. Phys. Rev. B **77**(12), 125401 (2008)
102. Koukaras, E.N., Kalosakas, G., Galiotis, C., Papagelis, K.: Phonon properties of graphene derived from molecular dynamics simulations. Sci. Rep. **5**, 12923 (2015)
103. Karssemeijer, L.J., Fasolino, A.: Phonons of graphene and graphitic materials derived from the empirical potential LCBOPII. Surf. Sci. **605**(17), 1611–1615 (2011)
104. Lindsay, L., Broido, D.A.: Optimized Tersoff and Brenner empirical potential parameters for lattice dynamics and phonon thermal transport in carbon nanotubes and graphene. Phys. Rev. B **81**(20), 205441 (2010)
105. Tersoff, J.: Empirical interatomic potential for carbon, with applications to amorphous carbon. Phys. Rev. Lett. **61**(25), 2879 (1988)
106. Mortazavi, B., Rajabpour A., Ahzi S., Hadizadeh kheirkhah, A.: Non-equilibrium molecular dynamics study on the thermal and mechanical properties of graphene. In: Proceedings of the 4th International Conference on Nanostructures (ICNS4), Kish Island, I.R. Iran, p. 1255, 12–14 Mar 2012
107. Magnin, Y., Förster, G.D., Rabilloud, F., Calvo, F., Zappelli, A., Bichara, C.: Thermal expansion of free-standing graphene: benchmarking semi-empirical potentials. J. Phys.: Condens. Matter **26**(18), 185401 (2014)
108. Lu, Q., Arroyo, M., Huang, R.: Elastic bending modulus of monolayer graphene. J. Phys. D Appl. Phys. **42**(10), 102002 (2009)

109. Huang, Y., Wu, J., Hwang, K.C.: Thickness of graphene and single-wall carbon nanotubes. Phys. Rev. B **74**(24), 245413 (2006)
110. Lu, Q., Huang, R.: Mechanical behavior of monolayer graphene by continuum and atomistic modeling. In: Simulations in Nanobiotechnology, p. 485. Taylor & Francis (2012)
111. Lampin, E., Priester, C., Krzeminski, C., Magaud, L.: Graphene buffer layer on Si-terminated SiC studied with an empirical interatomic potential. J. Appl. Phys. **107**(10), 103514 (2010)
112. Tang, C., Meng, L., Xiao, H., Zhong, J.: Growth of graphene structure on 6H-SiC (0001): Molecular dynamics simulation. J. Appl. Phys. **103**(6), 063505 (2008)

Perfecting the Growth and Transfer of Large Single-Crystal CVD Graphene: A Platform Material for Optoelectronic Applications

V. Miseikis, S. Xiang, S. Roddaro, S. Heun and C. Coletti

Abstract In this work, we demonstrate the synthesis of millimetre-sized single-crystals of graphene, achievable in a commercially available cold-wall CVD reactor, and several different approaches to transfer it from the growth substrate to a target substrate of choice. We confirm the high crystal quality of this material using various characterisation techniques, including optical and scanning electron microscopy as well as Raman spectroscopy. By performing field effect and quantum Hall effect measurements, we demonstrate that the electronic properties of such single crystals are comparable to those of ideal mechanically exfoliated flakes of graphene. Several applications of this high-quality material are also reviewed.

1 Introduction

Graphene grown by chemical vapour deposition (CVD) on copper (Cu) substrates has emerged as the preferable type of material for large-scale applications [1]. The low cost of growth substrates and the high electronic and structural quality of CVD graphene have ensured its widespread adoption in research. It has been shown, however, that a major limiting factor of the quality of CVD graphene are grain boundaries [2, 3]. Therefore, in recent years a major research focussing point has been to increase the crystal size of CVD graphene [4–7]. The key parameter for the growth of large-crystal graphene is the initial nucleation of graphene crystals on the highly reactive copper substrates. Various approaches have been proposed in order

V. Miseikis · C. Coletti (✉)
Center for Nanotechnology Innovation @NEST, Istituto Italiano di Tecnologia, Piazza San Silvestro 12, 56127 Pisa, Italy
e-mail: camilla.coletti@iit.it

S. Xiang · S. Roddaro · S. Heun
NEST, Istituto Nanoscienze—CNR and Scuola Normale Superiore, Piazza San Silvestro 12, 56127 Pisa, Italy

V. Morandi and L. Ottaviano (eds.), *GraphITA*, Carbon Nanostructures,
DOI 10.1007/978-3-319-58134-7_8

to lower the nucleation density, including performing the growth on highly isolated (enclosed) surfaces [4], strongly diluting the methane precursor [6], or oxidising the copper surface to passivate it [7]. It should be mentioned that the approaches listed above typically require either very long growth times [6], or extensive customisation of the CVD systems [7]. Recently, fast growth of inch-sized single crystals of graphene was reported on Cu/Ni alloys [8], although a sophisticated CVD system with local precursor supply was used in this case.

In this work, we present the development of a fast process for the synthesis of large-crystal graphene using a common, commercially available CVD system. We confirm the high structural and electronic quality of the transferred material by a variety of microscopy and spectroscopy techniques. Furthermore, we discuss the transfer of this material via different approaches: "wet" and "semi-dry" transfer. In fact, along with the charge carrier scattering by grain boundaries, transfer-induced contamination remains a limiting factor for the quality of CVD-grown graphene. The most common approach used for the transfer of graphene from copper and other catalysts involves chemical etching of the growth substrate using iron-based salts such as iron(III) chloride ($FeCl_3$) and iron(III) nitrate nonahydrate ($Fe(NO_3)_3$) [1, 9], or alternative "cleaner" etchants such as ammonium persulfate ($(NH_4)_2S_2O_8$ also known as APS) [5, 10]. More recently, the so-called "bubbling" transfer technique has been proposed, whereby the graphene is detached from the growth substrate via electrochemical delamination without the need to etch the metal [11, 12]. We report electric measurements displaying the remarkable quality of large single-grain graphene transferred using the electrochemical delamination in combination with a dry-handling technique. We finally discuss and present selected optoelectronic applications of CVD graphene.

2 Experimental

2.1 CVD Synthesis of Graphene

Large-crystal graphene was synthesised on commercially available copper foil with a thickness of 25 μm (99.8% purity, Alfa-Aesar, 13382), which prior to growth was electropolished as described previously [13] in order to remove the surface contamination and improve the surface flatness. This foil is also referred to later in the text as low-purity foil (i.e., with a high oxygen content). Several growth experiments were also performed with higher purity foil (99.98%, Sigma-Aldrich) with equivalent surface treatment.

The synthesis was performed using an *Aixtron BM Pro* cold-wall reactor equipped with a 10 cm heater and dual-heating set-up (top and bottom). All growth experiments presented here were performed at a pressure of 25 mbar. A typical

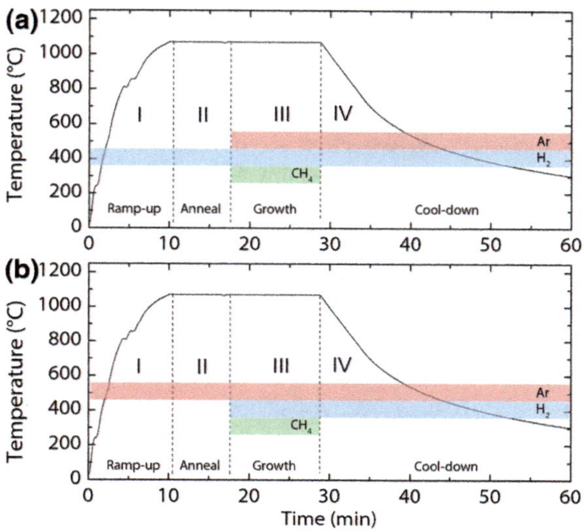

Fig. 1 Process diagram indicating the temperature profile and gas flow for the process using **a** hydrogen annealing and **b** argon annealing

growth procedure consisted of four distinct parts: temperature ramp-up, annealing, growth and cool-down (Fig. 1). A typical growth temperature was ~1060 °C. Temperature ramp-up (100 °C per minute) and annealing (duration: 10 min) was performed either in pure hydrogen (H_2) or pure argon (Ar) atmosphere, as will be discussed below. The growth (duration: 10–180 min, depending on the desired crystal size) was performed under a mixture of gas species (typical values of gas flow: 980 sccm Ar, 20 sccm H_2 and 1 sccm CH_4). The cool-down was performed in Ar/H_2 atmosphere (980 sccm/20 sccm). To avoid accelerated oxidation of copper, the samples were unloaded from the CVD reactor at a temperature below 120 °C.

2.2 *Transfer*

Two approaches were used to transfer graphene to the chosen substrates: "wet" and "semi-dry" transfer. "Wet" transfer was performed by first spin-coating the Cu/graphene sample with a PMMA support layer. Then the copper substrate was etched using an appropriate etchant (0.1 M $FeCl_3$ or 0.44 M (100 g/L) APS, Sigma-Aldrich), leaving the graphene/PMMA membrane floating on the surface of the etchant solution. The membrane was transferred to several baths of deionised water in order to rinse it thoroughly. The target substrate was then used to pick up the membrane from the deionised water, and the stack was dried in ambient conditions. Finally, the PMMA film was removed in acetone and isopropanol (IPA).

For the "semi-dry" transfer, the samples were initially spin coated with a PMMA support layer. A semi-rigid support frame made using polyimide adhesive tape (3M) was attached around the perimeter of the foil. To minimise contamination, chemical etching of the Cu substrate was not performed; instead, the graphene/PMMA stack was removed from the growth substrate by electrochemical delamination, using sodium hydroxide (NaOH) as electrolyte. The graphene/PMMA membrane (suspended and stretched flat by the support frame) was rinsed in deionised water, dried and deposited on the target substrate. To improve the adhesion, the samples were heated 10 min at 120 °C. Finally, the PMMA was removed in acetone and IPA.

2.3 Characterisation

The samples were analysed using a combination of microscopy and spectroscopy techniques. SEM imaging was performed using a *Zeiss Merlin* column with an accelerating voltage of 5 kV. Raman spectroscopy was performed using a *Renishaw inVia* system equipped with a 532 nm laser and a motorised stage for 2D mapping of samples.

The transport properties of large single-crystal graphene samples were investigated by conducting field effect and quantum Hall effect measurements. Isolated crystals were transferred to pre-patterned substrates of highly n-doped silicon with a 300 nm layer of SiO_2, which was chosen to allow back-gating of graphene devices. Subsequently, electron-beam lithography was used to contact the individual crystals in a six-terminal configuration. The measurements were performed using a lock-in technique, utilising a Heliox Helium-3 cryostat with base temperature of 250 mK.

3 Results and Discussion

3.1 Hydrogen Annealing Versus Argon Annealing

During the initial growth experiments, the ramp-up and annealing steps were performed in a pure hydrogen atmosphere. At high temperatures hydrogen is a strong reducing agent, helping to clean the Cu substrates from oxides and other impurities. This is a commonly used approach for the growth of polycrystalline films, however, due to a high nucleation density, it is unsuitable for synthesising large crystals of graphene. Growth experiments performed on copper foil with both low and high oxygen content yielded similar results (Fig. 2b, e, respectively) confirming nucleation densities as high as 10 000 per cm^2.

Recently it has been demonstrated that the presence of oxygen in the Cu substrate has a strong effect on the nucleation density and growth dynamics of CVD

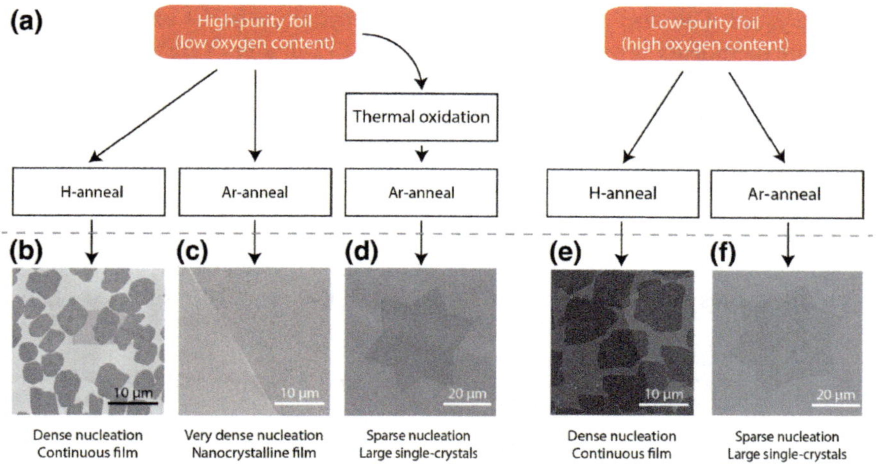

Fig. 2 **a** Process diagram of growth experiments performed with two kinds of copper foil. **b–f** SEM images of partially grown samples employing the different substrates and growth conditions. Note the different scale bars used for large-crystal images

graphene, making oxidised substrates desirable for large-crystal growth [7]. Due to safety concerns of introducing oxygen into a system containing highly combustible gases, as was done in the work cited above, we instead utilised the native oxides present in the low-purity Cu substrates [13]. To prevent the reduction of these oxides, annealing was performed in an inert argon atmosphere instead of hydrogen. Growth experiments performed using low-purity foil annealed in argon revealed a decreased nucleation density in the order of ~ 1000 grains per mm^2, an improvement of one order of magnitude. Additionally, annealing in argon provided an increased copper grain size of several millimetres.

In order to confirm that the presence of oxygen is the main factor limiting the nucleation density, Ar-annealing experiments were also performed using substrates with low oxygen content. As expected, this did not yield low nucleation density—on the contrary, the surface of such samples was often observed to be highly reactive, causing dense and uncontrollable nucleation and subsequent formation of nanocrystalline films (Fig. 2c). However, by simply heating the low-oxygen foil for 2 min at 180 °C in ambient atmosphere, the surface of such substrates could be intentionally oxidised. Subsequent growth experiments on such thermally oxidised foil provided qualitatively similar results to those obtained on natively oxygen-rich foil (Fig. 2d).

To further reduce the gas impingement flux on the sample, the foil (either low purity or thermally oxidised high purity) was contained within a small-volume custom-made enclosure, consisting of a quartz disk placed on 6 mm-thick graphite spacers. Such sample containment provided a significant further reduction of nucleation density to approximately 15 grains per mm^2. With such nucleation density, the growth time could be extended up to 1 h while retaining the isolated large crystals with size of up to 750 μm. In general, statistical analysis of many growth experiments revealed a growth rate of ~15 μm/min (Fig. 3b), among the highest reported to date. It is noteworthy that the use of an enclosure was beneficial only in combination with non-reducing argon annealing and did not have a strong effect of lowering the nucleation density on the highly reactive surface of Cu foil obtained using the hydrogen-annealing process.

To further decrease the nucleation density and allow synthesis of millimetre-sized single-crystals, the foils were folded into a pocket-like enclosure, initially proposed by Li et al. [4] and shown in Fig. 4a. Due to the crimped edges of such enclosures with a practically airtight seal, the inside Cu surfaces of these

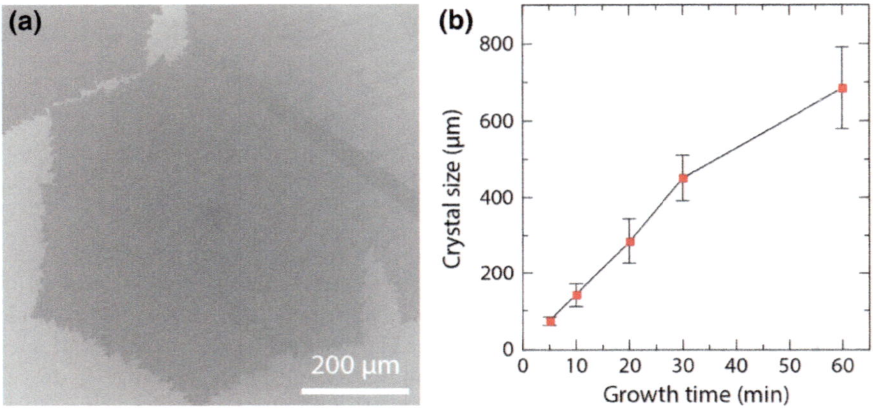

Fig. 3 **a** Single-crystal obtained after 1 h of growth. **b** Crystal size dependence on the growth time [13]

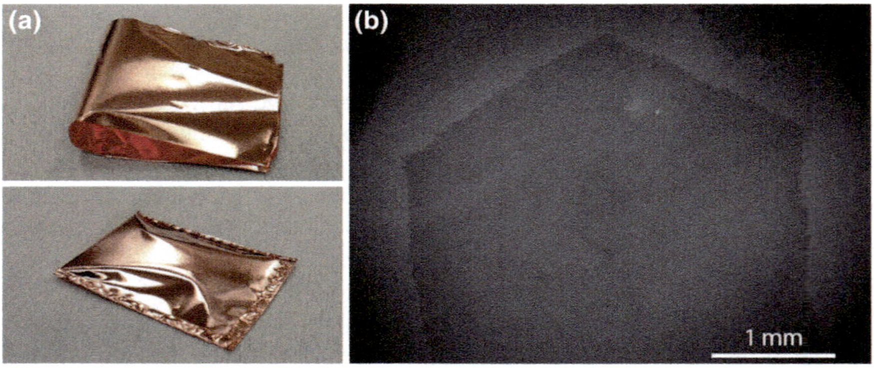

Fig. 4 **a** Formation of the copper "pocket": the foil is folded over (*top*), and the edges are crimped to create an airtight enclosure (*bottom*). **b** Single-crystal of graphene with a diameter of more than 3 mm grown inside the "pocket" enclosure [13]

"pockets" have little exposure to the carbon precursor (mainly due to the diffusion of carbon species through the copper [14]). This leads to extremely low nucleation densities of well below one crystal per mm^2, allowing to extend the growth time up to 3 h and to synthesise isolated single crystals of up to 3.5 mm (Fig. 4b). In the following sections, we discuss the transfer, properties, and applications of large single-grain graphene obtained in enclosed and oxidised low-purity Cu foil.

3.2 Optimisation of Graphene Transfer for Contamination-Free Graphene

As described in the experimental section, wet etching with iron(III) chloride and APS as well as electrochemical delamination were employed for the transfer of large grain graphene on various substrates. It should be noted that in the case of electrochemical delamination, we used a modified approach utilising a semi-rigid frame for dry handling of graphene/PMMA membranes. Notably, adoption of a frame allows attachment of the membrane to a micromanipulator stage for precise (in the order of a few μm) alignment of graphene single crystals on the target substrate.

Aside from microscopy techniques, a quantitative measure of the quality and cleanliness of transferred graphene is the analysis of the 2D peak in the Raman spectrum of graphene [15, 16]. In particular, the width of the 2D peak $\boldsymbol{\Gamma}_{(2D)}$ is an indicator of strain variations within the laser probe spot, which can also be responsible for charge carrier scattering, and hence is a major limitation of electronic mobility. Comparing the Raman spectra obtained from our samples prepared using the different transfer techniques, clear differences in the width of the 2D peak were consistently observed, even though the defect-related D peak was rarely observed in any of the samples, indicating the overall high crystal quality of graphene. $\boldsymbol{\Gamma}_{(2D)}$ after $FeCl_3$ transfer was typically found to be $\sim 35\ cm^{-1}$. When using APS, the typical $\boldsymbol{\Gamma}_{(2D)}$ values were $\sim 29\ cm^{-1}$. For the samples prepared using "semi-dry" transfer of electrochemically delaminated graphene, a typical $\boldsymbol{\Gamma}_{(2D)}$ value was $27\ cm^{-1}$, comparable to that of mechanically exfoliated flakes. Typical Raman spectra obtained in each case are shown in Fig. 5.

The "semi-dry" transfer undoubtedly provided the lowest impurity levels, thanks to the self-cleaning effect of the hydrophobic membranes. However, some samples transferred with this approach suffered adhesion issues. In particular, extra-large grains with a diameter above 500 μm and continuous films of graphene typically contained numerous pinholes. The formation of such defects could be explained by the trapping of small air pockets between the substrate and the graphene membrane, which is known to be impermeable to gas [17]. When the samples with freshly transferred graphene are placed in acetone to remove the PMMA support, the trapped air can escape by breaking the graphene and causing a pinhole. For smaller grains, the statistical probability of trapping air was much lower, and far fewer

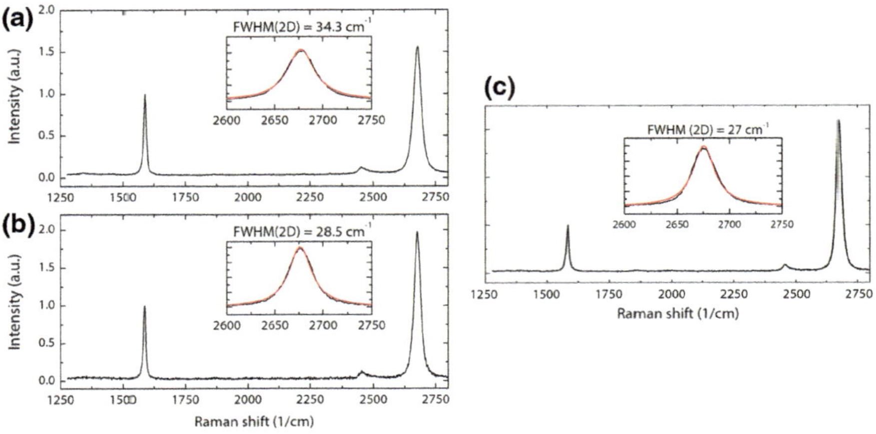

Fig. 5 Typical Raman spectra obtained from samples prepared using different transfer approaches. **a** Cu etching by $FeCl_3$, **b** Cu etching by APS, **c** electrochemical delamination and semi-dry transfer. Inset of each graph shows a Lorentz fitting of the 2D peak

damaged crystals were observed. On the other hand, during "wet" transfer, a thin and bubble-free layer of water was typically present at the graphene/substrate interface, which was subsequently slowly removed by the capillary effect, ensuring homogeneous adhesion of graphene. For this reason, when large graphene grains (with a diameter above 300 μm) need to be transferred, the wet transfer approach remains the preferable option, despite the relatively higher amount of contamination. Further work is needed to optimise the "semi-dry" transfer method and improve the adhesion of graphene on target substrates over large areas.

3.3 *Measurement of Transport Properties*

Transport measurements were conducted in graphene single-crystals transferred with the "semi-dry" approach and contacted in a six-terminal configuration, as shown in Fig. 6. A source-drain current of 10 nA was applied across the graphene crystal, and the side contacts were used to measure the longitudinal voltage drop V_{xx} and the transverse (Hall) voltage drop V_{xy}.

Electric field effect was measured by sweeping the back-gate voltage in the range of −45 to 60 V. A clear and sharp resistivity peak was observed at a back-gate voltage of around +8.5 V, indicating an impurity-induced charge density of approximately 6×10^{11} cm^{-2}, a relatively low value for graphene on Si/SiO_2. Furthermore, fitting the field effect measurement data using the Drude model (Fig. 7a) suggests carrier mobility above 10 000 cm^2/Vs, which is comparable to samples prepared on SiO_2 using pristine, mechanically exfoliated flakes.

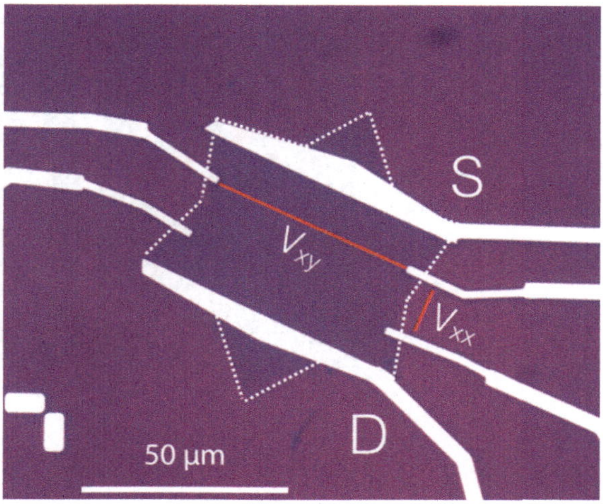

Fig. 6 Optical image of a large crystal of graphene contacted for transport measurements

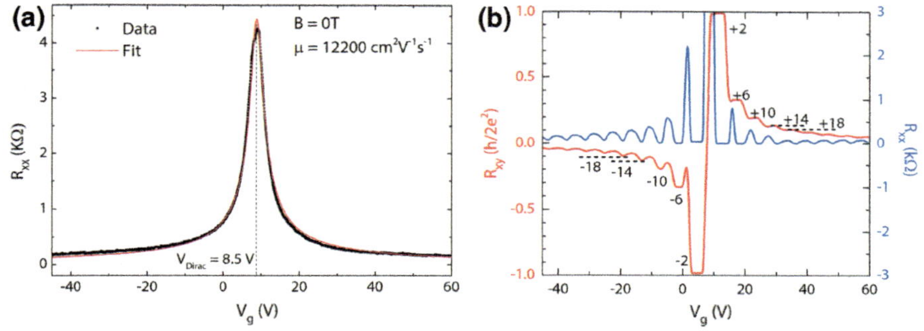

Fig. 7 **a** Electric field effect at zero applied magnetic field. **b** Traces of longitudinal resistance R_{xx} (*blue*) and transverse (Hall) resistance R_{xy} (*red*) as a function of V_g obtained at B = 4T

Quantum Hall measurements were performed by applying a magnetic field of up to 10 T. Remarkably, even though a non-ideal Si/SiO_2 substrate was used, upon application of magnetic field, pronounced oscillations of V_{xx} and plateaus of V_{xy} were observed, clearly indicating the quantum Hall effect. At a field of 4 T, up to 10 well-defined levels could be distinguished (Fig. 7b), further confirming the exceptionally high quality of the samples. A more thorough representation of the quantum Hall measurement of these samples is shown in the form of a Landau fan diagram (Fig. 8).

Furthermore, quantum interference was studied in these samples [18], by analysing the weak localisation peak at various back-gate voltages and temperatures, revealing a carrier dephasing length comparable to that of exfoliated graphene and larger than that previously observed in CVD-grown graphene.

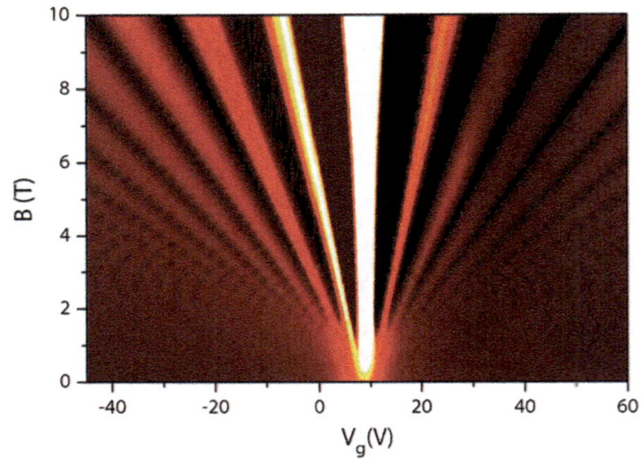

Fig. 8 Landau fan diagram of magnetoresistance R_{xx} as a function of V_g and B

The field effect and quantum Hall effect measurements presented above clearly demonstrate that the electric quality of single-crystal CVD graphene is comparable to that of mechanically exfoliated flakes. It should be noted that the measured mobility values were likely to be limited by the non-ideal Si/SiO_2 substrate. Previous work suggests that further improvements in transport characteristics of such samples of CVD graphene can be achieved by using high-quality substrates such as hexagonal boron nitride [5, 16].

4 Applications

The high quality of large single-crystal CVD graphene has made it attractive for various applications. Recently, Spirito et al. used this material to fabricate UV-sensitive photodetectors [19]. The devices contained graphene field effect transistors coated with a sensitising layer of colloidal CdS nanocrystals—i.e., the nanocrystals absorbed the incident light creating electron–hole pairs; and the electrons were then transferred to graphene, generating a detectable photocurrent. The detectors demonstrated high responsivity of up to 10^4 A/W and a fast response rate in the kHz range.

The high quality of graphene was even more important in a recent investigation of transport in strongly coupled graphene and $LaAlO_3/SrTiO_3$ oxide (LAO-STO) junctions [20]. In order to maintain a pristine, contamination-free interface between the graphene single-crystals and LAO-STO, the "semi-dry" transfer technique was employed during the fabrication of these samples. At low applied bias between the graphene and LAO-STO, strong interlayer electrostatic coupling was observed, allowing the gating of graphene. On the other hand, at higher vertical bias above |1 V|, direct tunnelling coupling was measured.

As discussed above, large crystals of CVD graphene can demonstrate electric properties comparable to those measured using mechanically exfoliated flakes. However, in certain cases where the high mobility of graphene is not a necessary prerequisite, polycrystalline films prepared using the hydrogen annealing could provide significantly larger continuous areas of graphene than are possible to achieve with single-crystals. One application of such material was a hybrid metasurface consisting of terahertz split-ring resonator (SRR) arrays and graphene, investigated by Zanotto et al. [21]. Wet transfer was used to deposit a sheet of graphene on GaAs substrates containing a gold split-ring resonator metamaterial and a 30 nm insulating layer of SiO_2. The presence of graphene modified the THz transmission response of the metamaterial allowing its modulation by an applied magnetic field. Notably, a magnetic field of 5 T applied to a hybrid graphene/SRR metasurface caused transmittance modulation of up to 10%, compared to just $\sim 1\%$ modulation of unpatterned graphene/SiO_2/GaAs stack, indicating a significant interaction between the Dirac fermions in graphene and the SRR resonance. Furthermore, numerical simulations of magneto-optic response of such hybrid material were performed, with good agreement to the experimental data.

5 Conclusions

We have presented a route to obtain high-quality single-crystal graphene using a commercially available CVD system. Using a variety of characterisation techniques, we confirm the high quality of the material, and we present several approaches to integrate this material on different substrates of choice. We demonstrate that using state-of-the-art transfer techniques, it is possible to fabricate large-area devices with electric quality comparable to that of mechanically exfoliated graphene. Finally, we present some recent applications of our material.

Acknowledgements We acknowledge funding from the European Union Seventh Framework Programme under grant agreement no. 604391 Graphene Flagship. Part of the work received financial support from the Italian Ministry of Foreign Affairs (Ministero degli Affari Esteri, Direzione Generale per la Promozione del Sistema Paese) in the framework of the agreement on scientific collaborations with Canada (Quebec) and Poland; and from the CNR in the framework of the agreement on scientific collaborations between CNR and JSPS (Japan), CNRS (France), and RFBR (Russia).

References

1. Li, X., et al.: Large-area synthesis of high-quality and uniform graphene films on copper foils. Science **324**, 1312–1314 (2009)
2. Yu, Q., et al.: Control and characterization of individual grains and grain boundaries in graphene grown by chemical vapour deposition. Nat. Mater. **10**, 443–449 (2011)

3. Zhang, H., Lee, G., Gong, C., Colombo, L., Cho, K.: Grain boundary effect on electrical transport properties of graphene. J. Phys. Chem. C **118**, 2338–2343 (2014)
4. Li, X., et al.: Large-area graphene single crystals grown by low-pressure chemical vapor deposition of methane on copper. J. Am. Chem. Soc. **133**, 2816–2819 (2011)
5. Petrone, N., et al.: Chemical vapor deposition-derived graphene with electrical performance of exfoliated graphene. Nano Lett. **12**, 2751–2756 (2012)
6. Zhou, H., et al.: Chemical vapour deposition growth of large single crystals of monolayer and bilayer graphene. Nat. Commun. **4**, 2096 (2013)
7. Hao, Y., et al.: The role of surface oxygen in the growth of large single-crystal graphene on copper. Science **342**, 720–723 (2013)
8. Wu, T., et al.: Fast growth of inch-sized single-crystalline graphene from a controlled single nucleus on Cu-Ni alloys. Nat. Mater. **15**, 43–47 (2015)
9. Kim, K.S., et al.: Large-scale pattern growth of graphene films for stretchable transparent electrodes. Nature **457**, 706–710 (2009)
10. Bae, S., et al.: Roll-to-roll production of 30-inch graphene films for transparent electrodes. Nat. Nanotechnol. **5**, 1–5 (2010)
11. Wang, Y., et al.: Electrochemical delamination of CVD-grown graphene film: toward the recyclable use of copper catalyst. ACS Nano **5**, 9927–9933 (2011)
12. Gao, L., et al.: Repeated growth and bubbling transfer of graphene with millimetre-size single-crystal grains using platinum. Nat. Commun. **3**, 699 (2012)
13. Miseikis, V., et al.: Rapid CVD growth of millimetre-sized single crystal graphene using a cold-wall reactor. 2D Mater. **2,** 014006 (2015)
14. Fang, W., et al.: Asymmetric growth of bilayer graphene on copper enclosures using low-pressure chemical vapor deposition. ACS Nano **8**, 6491–6499 (2014)
15. Neumann, C., et al.: Raman spectroscopy as probe of nanometre-scale strain variations in graphene. Nat. Commun. **6**, 8429 (2015)
16. Banszerus, L., et al.: Ultrahigh-mobility graphene devices from chemical vapor deposition on reusable copper. Sci. Adv. **1,** e1500222–e1500222 (2015)
17. Bunch, J.S., et al.: Impermeable atomic membranes from graphene sheets. Nano Lett. **8**, 2458–2462 (2008)
18. Xiang, S. et al.: Low-temperature quantum transport in CVD-grown single crystal graphene. Nano Res. **9**, 1823 (2016)
19. Spirito, D., et al.: UV light detection from CdS nanocrystal sensitized graphene photodetectors at kHz frequencies. J. Phys. Chem. C **119**, 23859–23864 (2015)
20. Aliaj, I. et al.: Tunnel and electrostatic coupling in graphene-LaAlO3/SrTiO3 hybrid systems, APL Mat **4**, 066101 (2016)
21. Zanotto, S., et al.: Magneto-optic transmittance modulation observed in a hybrid graphene–split ring resonator terahertz metasurface. Appl. Phys. Lett. **107**, 121104 (2015)

Advances in the Fabrication of Large-Area Back-Gated Graphene Field-Effect Transistors on Plastics: Platform for Flexible Electronics and Sensing

G. Fisichella, S. Lo Verso, S. Di Marco, V. Vinciguerra, E. Schilirò, S. Di Franco, R. Lo Nigro, F. Roccaforte, A. Zurutuza, A. Centeno, S. Ravesi and F. Giannazzo

Abstract Graphene (Gr) is currently one of the most appealing materials as conductive transparent electrode for flexible electronics, thanks to its bendability/stretchability accompanied by small variations of the electrical properties after mechanical deformations. In addition, the field-effect tunable carrier density combined to a high mobility and saturation velocity make it an excellent channel material for field-effect transistors (FETs) even on flexible substrates. By proper design of the device structure (channel length, top- or back-gate configuration), Gr-FETs can be used for high-frequency (RF) electronics or for high-sensitivity chemical, biological, and environmental sensors exploiting transconductance variations in response to the chemi/physisorption of molecular species on Gr channel. In particular, miniaturized and flexible Gr-FET sensors can represent a strong advance with respect to current sensors technology and will be extremely useful for "in situ" applications. Here we report a wafer scale and semiconductor fab compatible processing strategy to fabricate arrays of Gr-FETs on a PEN substrate, adopting a local back-gate configuration, with a thin Al_2O_3 gate dielectric film deposited at low temperature (100 °C) by plasma-assisted Atomic Layer Deposition (ALD) and transfer of large-area Gr grown by chemical vapor deposition on copper foils. Electrical characterization of the fabricated devices is presented and their suitability for solid ion sensing FET (IS-FET) applications is discussed.

G. Fisichella · E. Schilirò · S. Di Franco · R. Lo Nigro · F. Roccaforte · F. Giannazzo (✉)
CNR-IMM, VIII Strada, 95121 Catania, Italy
e-mail: filippo.giannazzo@imm.cnr.it

S. Lo Verso · S. Di Marco · V. Vinciguerra · S. Ravesi
STMicroelectronics, Stradale Primosole 50, 95121 Catania, Italy

A. Zurutuza · A. Centeno
Graphenea, Tolosa Hiribidea 76, Donostia-San Sebastian, Spain

V. Morandi and L. Ottaviano (eds.), *GraphITA*, Carbon Nanostructures,
DOI 10.1007/978-3-319-58134-7_9

1 Introduction

Recently the flexible device technology has turned out as one of the new challenges in the field of electronics. From stretchable, foldable, or form factor reconfigurable devices [1] to biomedical in vivo applications [2, 3], the recent growing interest on flexible electronics has been spurred by the evolution of the processing technologies for soft substrates and by the discovering of new materials suitable for bending, stretching or conformably shaping [4], with a particular attention for the 2D materials [5]. Among them, graphene (Gr) is one of the most appealing as conductive membrane capable of bearing deformations, exhibiting a Young's modulus in the order of TPa and a large spring constant (1–5 N/m) [6]. Another fundamental property which makes Gr the natural choice for flexible electronics is represented by its high-charge carriers mobility, that can reach thousands of $cm^2V^{-1}s^{-1}$ even at room temperature for both electrons and holes, which is a 2–3 orders of magnitude higher value compared to the other candidates considered for flexible applications, such as metal oxides, amorphous Si, or organic semiconductors. It should be also noted that these properties are accompanied by small variations of the electrical performances under mechanical deformations [7, 8], which is an essential requirement for devices operation stability and reliability. The high mobility allowed Gr exploitation for the fabrication of high frequency Gr-FETs not only on conventional substrates [9], but also on flexible plastics ones [10]. In particular, remarkable cut-off frequency values (~25 GHz) and robust performances under repeated bending (down to 0.7 mm bending radius) have been recently obtained in Gr-FETs even with not aggressively scaled (~0.5 μm) channel lengths fabricated on a flexible polyimide substrate, using a back-gate configuration and Al_2O_3 as a gate dielectric.

Another field where Gr can be profitably exploited is chemical/biological sensing. Due to the low density of states (DOS) around the Dirac point, Gr carrier density is very sensitive to the adsorption of charged/polar species. Gr-FET based sensors can represent a great combination between a chemical-to-electrical signal converter and an electrical signal amplifier [11]. The key electrical parameter ruling the electrical response of a Gr-FET is the transconductance g_m, which is defined as

$$g_m = \frac{\mu C_g W V_{DS}}{L}, \quad (1)$$

where μ is the channel mobility, C_g is the gate capacitance, V_{DS} is the drain bias, L and W are the channel length and width, respectively. As a matter of fact, large-area channels are required for sensing applications. Hence, a high gate capacitance in combination with the high Gr mobility is necessary to maximize the Gr-FET sensor transconductance.

Among the different kinds of sensor configurations, the ion sensing FET (IS–FET) [12] is one of the most used for chemical and biological sensing. Recently it has been implemented also with Gr. In this case the Gr channel is covered by the target solution and a macroscopic reference electrode immersed in the solution itself works as a gate contact, as schematically reported in Fig. 1a. In particular, the electric double layer capacitances at the gate/solution interface and at the channel/solution interface play the major role for the IS–FET capacitance which can reach the order of magnitude of tens μF/cm^2 (depending on the ions concentration). It worth noting that these capacitances values are in the same order of magnitude of the Gr quantum capacitance [13, 14], which is ranging from ideally zero to tens of μF/cm^2, depending on the channel electrostatic potential. The need of a reference electrode, which is an external, rigid, and macroscopic element, enormously limits conventional IS-FETs both for autonomous device applications and for positioning in complex locations. A different device configuration, which can be more efficiently subjected to miniaturization and integration, is the solid IS–FET [15], where the macroscopic reference electrode is replaced by a local back-gate buried under a dielectric film (Fig. 1b). In this case the thickness and the dielectric constant of the insulating film have a crucial importance in order to maintain a reasonably high transconductance of the final devices. In particular, considering high κ-dielectrics such as HfO_2 or Al_2O_3 with film thickness in the order of 10 nm, the gate capacitance is in the order of the unit of μF/cm^2. In this sense, losing about an order of magnitude in the gate capacitance (i.e., in the transconductance) is compensated by the advantages in terms of devices miniaturization and integration. Considering properly thin high-k dielectrics it is still possible to detect the Gr quantum capacitance as relevant series contribution for the whole total capacitance. It should be added that the high capacitance per unit area of such kind of devices can also allow the use of a metal-insulator-graphene capacitor (MIGr) configurations as a variable capacity sensing device [16] while the high mobility in Gr would enable high quality factor (Q) to be achieved [17].

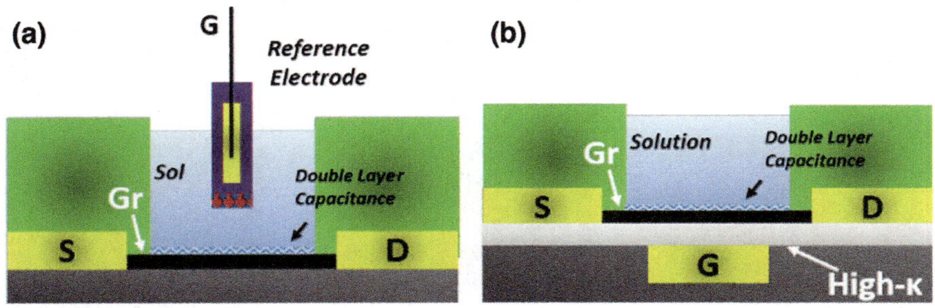

Fig. 1 Schematic representation of the IS-FET with **a** a solution gate configuration and **b** a solid state configuration

2 Transistors Fabrication

Here we report a silicon fab compatible fabrication flow to obtain wafer scale arrays of independently back-gated Gr-FETs. In particular, we developed the fabrication of large-area devices with channel widths/lengths in the order of magnitude of 10^2 µm. Such widely extended channels are particularly suitable for solution sensing applications where the sensitive surface must be able to efficiently interface the target solution. However, these sizes pose a challenge considering that the bigger is the channel dimension the higher is the effect on the single device performances due to the materials defects (graphene cracks and grain boundaries, surface asperities, dielectric inhomogeneity).

The plastic substrate adopted in the following is a thermally flattened polyethylene naphthalate (PEN) Teonex film purchased by Dupont Teiji Films. PEN is a polymer analogous of the more commonly used polyethylene terephthalate (PET), but with superior physical properties and thermal stability (up to 150 °C), which represents an optimal choice as substrate for flexible technology.

In order to perform all the device processing steps within industrial equipments commonly used for silicon devices fabrication, the polymer film was reversibly bonded on a 150 mm Si wafer by mechanical lamination, adopting a double face thermal release tape (150 °C thermal release) and shaped by cutting along the wafer edges (Fig. 2a).

The PEN substrate bonded on wafer, as reported in Fig. 2b, was morphologically characterized by tapping mode Atomic Force Microscopy (tAFM) (Fig. 2c) and a starting root means square (RMS) of ~7 nm was evaluated. It is worth noting

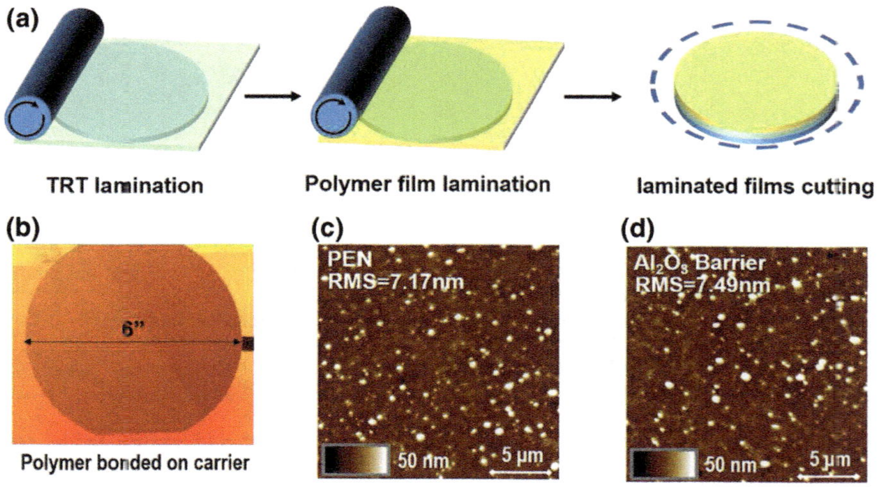

Fig. 2 **a** Schematic representation of the reversible bonding of PEN on silicon wafer by lamination, **b** photograph of the PEN substrate bonded on wafer, **c** tAFM morphology of the PEN surface and **d** tAFM morphology of PEN coated by a 100 nm protective Al_2O_3 layer

that, even if this is a reasonable roughness for a plastic substrate, it is 1–2 orders of magnitude higher than the RMS values of traditional rigid substrates for electronics, such as a silicon dioxide on Si.

In order to protect the polymer film surface from degradation due to the processing steps (in particular where plasma is involved), a 100 nm protective Al_2O_3 coating was deposited by DC pulsed RF reactive sputter, assisted by a cooling system able to maintain the sample below 100 °C. The tAFM morphology reported in Fig. 2d shows an essentially unchanged morphology of the substrate surface after the protective coating deposition, demonstrating that the deposition itself does not damage the PEN surface significantly.

200 nm thick aluminum back gate pads were fabricated by metal sputtering and lift-off, as reported in the schematic illustration in Fig. 3a. As reported in Fig. 3b, the tAFM morphology of the Al contact surface shows a roughness of ~10 nm, slightly higher than the underlying substrate roughness.

As schematically illustrated in Fig. 3c, a ~ 30 nm thick Al_2O_3 dielectric film was deposited by Atomic Layer Deposition (ALD). In particular, a properly optimized plasma assisted ALD process was adopted considering Trimethylaluminium (TMA) as the metalorganic precursor and plasma activated O_2 as coupled reagent for the Al_2O_3 formation. A low temperature process (T = 100 °C) was exploited in order to preserve the plastic substrate. Such a low temperature, compared to the typical deposition temperature of Al_2O_3 by ALD, which is in the range of 200–250 °C, required several extra expedients for a good-quality dielectric growth. In particular, a larger oxygen dose per growth cycle and prolonged purging periods are necessary to compensate the slower reactivity and the less-efficient TMA excess desorption due to the lower involved temperature. The surface tAFM morphology reported in Fig. 3d shows a slightly lower roughness of ~8 nm than the underlying

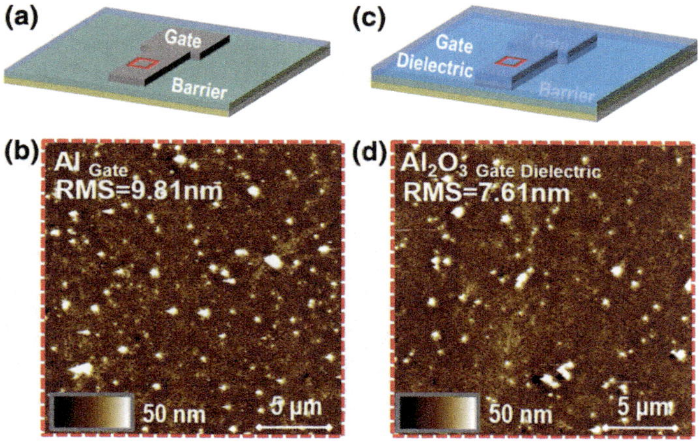

Fig. 3 **a** Schematic illustration of the aluminum back-gate and **b** related tAFM morphology. **c** Schematic illustration of the Al_2O_3 deposited by ALD as the back gate dielectric and **d** related tAFM morphology

Al gate contact. This effect can be explained considering the conformal coating of the underlying surface morphology with a consequent smoothing of the asperities profile and a lowering of the average roughness.

For the fabrication of the Gr-FET channels, we used single-layer Gr films grown by chemical vapor deposition (CVD) on large are copper foils provided by Graphenea. Figure 4a shows a photograph of circular copper foils (100 mm diameter) with as-grown Gr. The Gr membrane was transferred to the target substrate by a PMMA-assisted wet transfer procedure.

Gr starting quality was characterized transferring the material over a silicon reference substrate coated by 300 nm of thermally grown SiO_2. In this case, the lower substrate roughness allows a more effective morphological characterization while the Raman spectroscopy is not affected by the high background signal associated to the polymeric PEN substrate. Figure 4b shows the tAFM morphology of Gr transferred on SiO_2/Si that results continuous and defect-free, with a reasonable density of characteristics wrinkles and a low percentage per unit area (below 5%) of hexagonal second layer nucleated grains. Figure 4c illustrates the

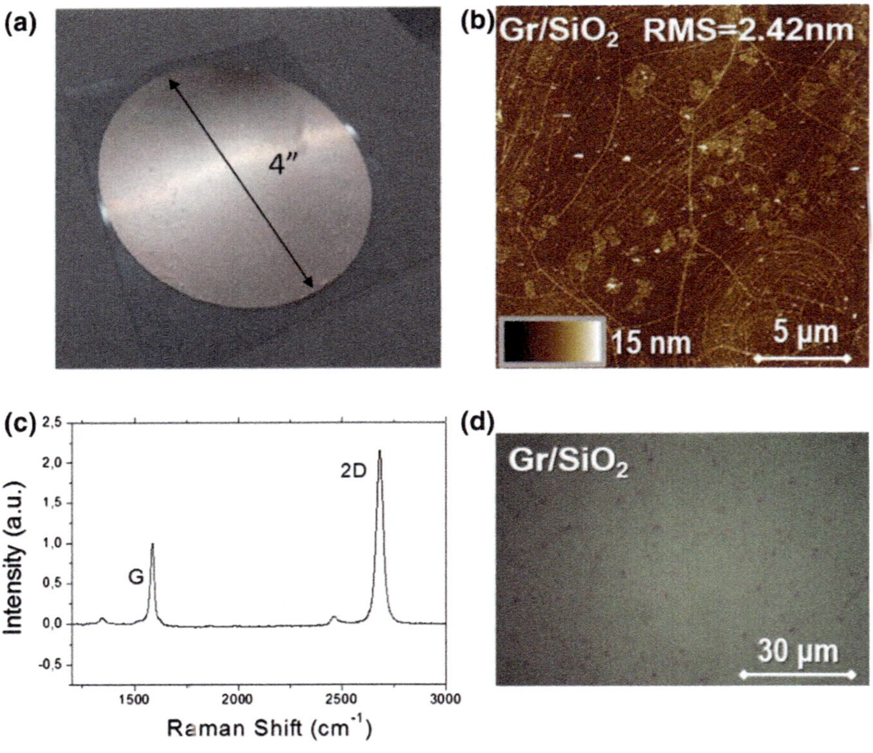

Fig. 4 **a** Photograph of a circular copper foil (100 mm diameter) with as-grown Gr, **b** tAFM morphology, **c** Raman spectra and **d** optical microscopy image of Gr transferred on SiO_2/Si

Raman spectra of the material which demonstrate the high Gr structural quality. In particular the negligible intensity of the D peak ($I_D/I_G < 0.05$) suggests a low level of structural defects while the ratio between the 2D and the G peaks intensity ($I_{2D}/I_G \approx 2.1$), and the symmetry of the 2D peak (FWHM ~ 40 cm^{-1}, Lorentzian fit) demonstrates the single-layer character of the Gr film. Finally, Fig. 4d shows the optical microscopy image of the transferred membrane which confirms the Gr continuity on a large area.

After transferring of the large area (100 mm diameter) Gr membrane on the 150 mm diameter PEN wafer with the back-gate structures, Gr channel regions were lithographically defined by O_2 plasma etching. Figure 5a shows an optical microscopy image of the device at this intermediate fabrication step, with the Gr channel region highlighted by the blue transparency. A magnification of the Gr channel edge is reported in Fig. 5b. Finally, the tAFM image in Fig. 5c shows the peculiar Gr morphology which is typically less evident on PEN substrates compared with smoother substrates. However, it is still possible to identify wrinkles, as indicated by the dashed green circles in Fig. 5c. The density and the average dimension and height of the surface asperities (white spots in the AFM image) are consistent with what reported on PEN surface before the transfer of Gr, indicating that they are mainly due to the substrate than to the transfer procedure of Gr.

FETs fabrication flow is completed by the fabrication of 30/120 nm thick Ni/Au double metal source and drain contacts, obtained by sputtering and lift-off. Figure 6a reports the optical microscopy image of a Gr-FET while Fig. 6b shows the whole view of the wafer, which includes arrays of differently shaped and sized back-gate transistors. Figure 6c reports the schematic illustration of a local back gate Gr-FET. In particular, the back gate defines a step topography and Gr follows the topography and is contacted by source and drain away from the channel region. For this device structure, the measured resistance between the source and drain

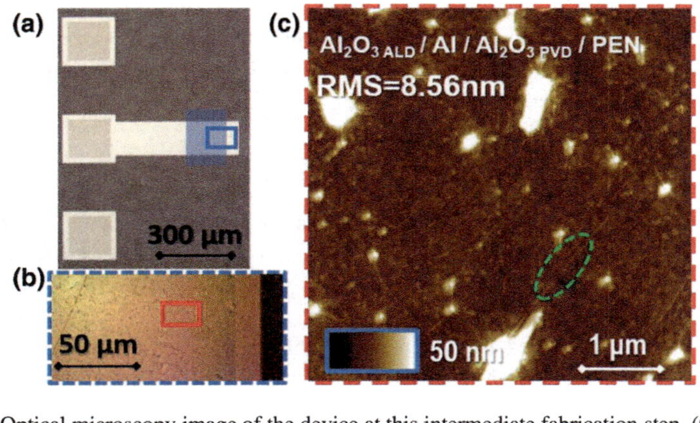

Fig. 5 **a** Optical microscopy image of the device at this intermediate fabrication step, (Gr channel region highlighted in *blue*), **b** magnification of the blue rectangle in (**a**), where the patterned Gr edge is visible and **c** tAFM morphology of Gr in the same channel region

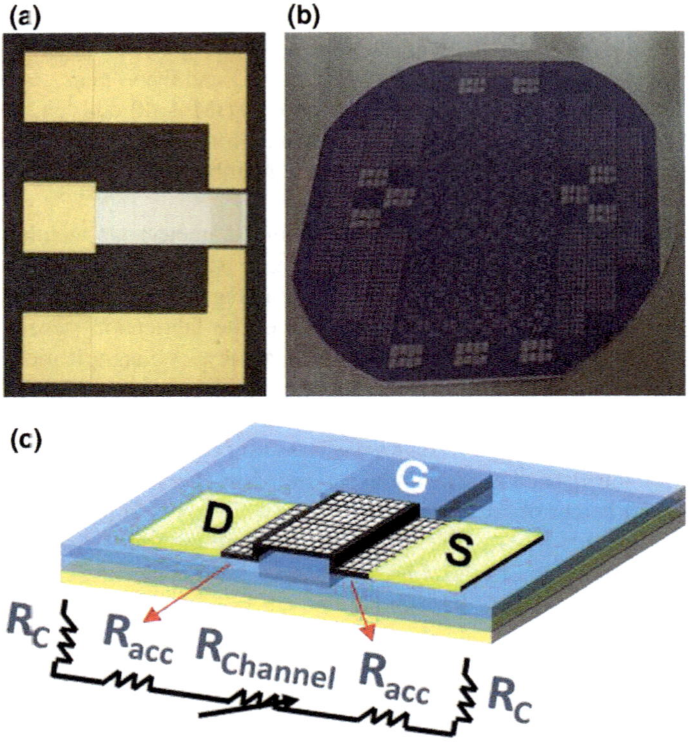

Fig. 6 **a** Optical microscopy image of a Gr-FET, **b** photograph of the wafer and **c** schematic 3D illustration of a local back gate Gr-FET with the detail of the series resistances contributions

contact (R_{TOT}) is the series combination of different resistive contributions, i.e., the gate-bias-dependent Gr channel resistance $R_{ch}(V_g)$, the contact resistance R_c between Gr and the metal source and drain contacts, and the access resistance R_{acc} associated to the ungated Gr access regions between the channel and the source and drain contacts with a length per access region $L_{acc} = 20$ μm.

$$R_{TOT}\left(V_g\right) = R_{ch}\left(V_g\right) + 2R_c + 2R_{acc}, \tag{2}$$

Before the Gr-FET electrical characterization, the capacitance of the Al_2O_3 dielectric deposited by ALD was characterized by Metal-Insulator-Metal (MIM) test devices properly manufactured in the same wafer. The evaluated gate capacitance per unit area is $C_g = 2.05 \times 10^{-7}$ F/cm^2. The Al_2O_3 film thickness (29.9 nm) was determined by spectroscopic ellipsometry on a reference Si wafer where an identical ALD deposition was performed. The resulting dielectric constant is 6.9, reasonably high considering the low temperature (100 °C) adopted for the dielectric growth. As a way of comparison with silicon dioxide, the resulting Equivalent Oxide Thickness (EOT) is 16.8 nm.

The typical ambipolar behavior with a minimum current at the neutrality point (V_{NP}) is observed. Clearly, the neutrality point is significantly shifted at high positive bias with respect to the expected ideal value calculated considering the difference between Al and neutral Gr workfunctions ($V_{NI,id} = W_{Al} - W_{Gr} \approx 4.1 - 4.5 = -0.4$ eV). This positive shift is a clear indication of Gr p-type doping, that can be estimated as $p = C_g(V_{NI} - V_{NI,id})/q \approx 1.2 \times 10^{13}\ \mathrm{cm}^{-2}$, being q the electron charge. This doping can be ascribed both to the effect of the chemical (PMMA) residues which normally persist after the Gr transferring and of the resist residues from lithographic steps, and, partially, to the electrostatic effect of fixed charges in the gate oxide deposited at low temperature. It is worth noting that removal of PMMA residues from Gr after transfer to common substrates (such as SiO_2/Si) is typically obtained by performing annealing processes in vacuum or Ar/H_2 ambient at temperatures from 300 to 400 °C. As a matter of fact, such cleaning processes cannot be adopted in the present case, due to the low-temperature process constrains (150 °C) imposed by the used plastic substrate.

The insert of Fig. 7a shows the Gr-FET output characteristics ($I_d - V_d$) for gate-bias values in the range between $V_g = 0$ V and $V_g = 11$ V, i.e., slightly above V_{Dirac}. Clearly, all the characteristics exhibit a linear behavior in the considered source-drain bias range (V_d from −0.9 to +0.9 V), as expected for long channel Gr-FETs.

Figure 7b shows the measured current through the gate dielectric (I_g), demonstrating that the leakage is below 1 nA almost across the whole considered bias range and completely negligible compared to the channel currents. This is a clear indication of the dielectric film quality achieved in spite of the low ALD process temperatures.

The total resistance R_{TOT} between the source and drain contacts as a function of V_g is calculated as $R_{TOT} = V_d/I_d$ and reported in Fig. 8a, back curve. As previously discussed (see Eq. 2), R_{TOT} is the combination of $R_{ch}(V_g)$ and of a

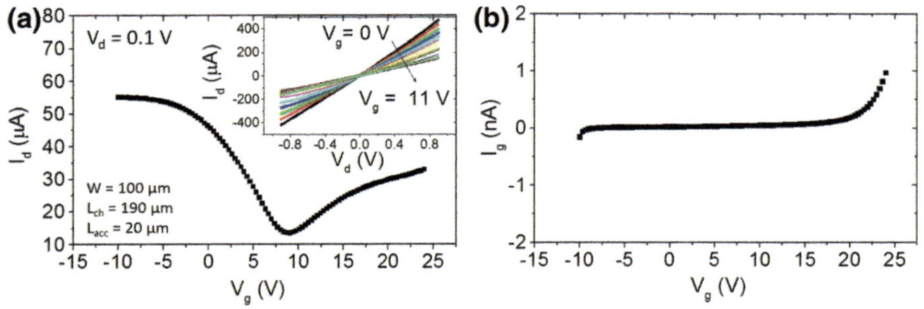

Fig. 7 **a** Transfer characteristic for a representative Gr-FET with channel width W = 100 μm and channel length L = 190 μm, with Gr-FET output characteristics ($I_d - V_d$) for gate-bias values in the range between $V_g = 0$ V and $V_g = 11$ V as an insert and **b** leakage current through the gate dielectric (I_g) in the same characterization range

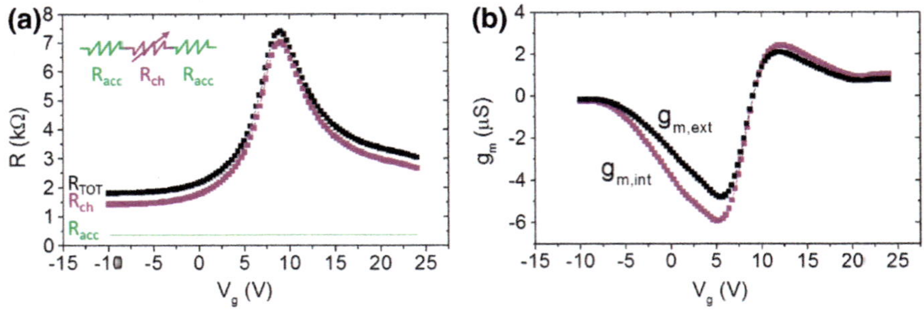

Fig. 8 **a** Total resistance R_{TOT} (*black curve*) and channel resistance R_{ch} (*magenta curve*) between the source and drain contacts as a function of V_g and the gate–bias-independent series resistance R_{acc} contribution (in *green*). **b** Extrinsic transconductance g_m of the Gr-FET, obtained by differentiation of the measured $I_d - V_g$ characteristic (*black curve*) and intrinsic transconductance corrected for the series resistance value (*magenta curve*)

gate-bias-independent series resistance contribution ($R_s = 2R_c + 2R_{acc}$). Such contribution has been evaluated by measuring the resistance between source and drain contacts at $V_g = 0$ in several transistors with the same W and L_{acc} and different channel length L. The estimated value of R_{acc} is also reported in Fig. 8a, green curve. The R_{ch} versus V_g curve, obtained subtracting R_s from R_{TOT}, is reported as magenta curve.

The transconductance g_m of the Gr-FET, obtained by differentiation of the measured $I_d - V_g$ characteristic ($g_m = dI_d/dV_d$), is reported in Fig. 8b, black curve. This is clearly the extrinsic transconductance, as it is affected by the above-mentioned series resistance contributions. The extrinsic hole and electron mobilities ($\mu_{h,extr} = 476\ cm^2V^{-1}s^{-1}$ and $\mu_{e,extr} = 204\ cm^2V^{-1}s^{-1}$) were evaluated from the peak transconductance values using to the formula $\mu = g_mL/(WC_gV_d)$. The intrinsic transconductance, corrected for the series resistance value reported in Fig. 8a, is also reported in Fig. 8b magenta curve. The corresponding intrinsic hole and electron mobility values are $\mu_{h,intr} = 581\ cm^2V^{-1}s^{-1}$ and $\mu_{e,intr} = 235\ cm^2V^{-1}s^{-1}$.

The obtained mobility values are reasonably high, considering the wide area of the channel, the high roughness due to the plastic substrate and the relatively high doping of Gr. In fact, due to the polycrystalline nature of CVD Gr, current transport in large-area channels is strongly affected by typical extended defects, such as grain boundaries. In addition to these natural defects originating from CVD growth, Gr membrane is subjected to significant strain if transferred to a rough surface, Furthermore, a certain amount of mechanical defects, such as cracks or folding can be introduced by the transfer procedure, especially onto a structured surface. Besides these macroscopic defects, nanoscale scattering mechanisms [18], such as charged impurities scattering and resonant scattering by defects/polymer contaminations, is expected to play a key role in reducing Gr mobility.

3 Conclusion

In conclusion, we reported recent advances in the fabrication of Gr-FET with large channel areas (~10^4 μm^2) and local back-gate on flexible PEN substrates using a silicon fab compatible fabrication flow. Various wafer-scale fabrication techniques where involved and properly redefined to effectively work at low temperatures compatible with the plastic substrates. In particular, a great effort was devoted to optimize deposition processes, such as DC pulsed reactive sputter of thick Al_2O_3 films working as protective barrier layer on PEN and plasma assisted atomic layer deposition at low temperature (100 °C) of thin (10–30 nm) and high quality Al_2O_3 working as back-gate dielectric of Gr-FET. Electrical characterizations of the Gr-FET devices are reported in order to evaluate key electrical parameters such-us the transconductance, Gr doping and electrons and holes mobility (235 $cm^2V^{-1}s^{-1}$ and 581 $cm^2V^{-1}s^{-1}$, respectively). The fabricated devices will represent the platform for the implementation of solid IS-FETs that can be part of a fully flexible integrated system for sensing and signal processing.

Acknowledgements The authors acknowledge Dr. P. Fiorenza from CNR-IMM, Catania, for useful discussions. This work has been supported, in part, by European Union Seventh Framework Program under Grant Agreement No. 604391 Graphene Flagship.

References

1. Kim, D.-H., Ahn, J.-H., Choi, W.M., Kim, H.-S., Kim, T.-H., Song, J., Huang, Y.Y., Liu, Z., Lu, C., Rogers, J.A.: Science **320**, 507 (2008)
2. Viventi, J., Kim, D., Moss, J.D., Kim, Y., Justin, A., Annetta, N., Hicks, A., Xiao, J., Huang, Y., David, J., Rogers, J.A., Litt, B.: Sci. Transl. Med. **2**, 24ra22 (2010)
3. Hwang, S.-W., Tao, H., Kim, D.-H., Cheng, H., Song, J.-K., Rill, E., Brenckle, M. A., Panilaitis, B., Won, S.M., Kim, Y.-S., Song, Y.M., Yu, K.J., Ameen, A., Li, R., Su, Y., Yang, M., Kaplan, D.L., Zakin, M.R., Slepian, M.J., Huang, Y., Omenetto, F.G., Rogers, J.A.: Science **337**, 1640 (2012)
4. Cheng, C., Wagner, S.: In: Salleo, A., Wong, W.S. (eds.) Flexible Electronics: Materials and Applications. Springer
5. Akinwande, D., Petrone, N., Hone, J.: Nat. Commun. **5**, 5678 (2014)
6. Lee, C., Wei, X., Kysar, J.W., Hone, J.: Science **321**, 385–388 (2008)
7. Kim, K.S., Zhao, Y., Jang, H., Lee, S.Y., Kim, J.M., et al.: Nature **457**, 706–710 (2009)
8. Kwak, B.W., Choi, Y.C., Lee, B.S.: Phys. E Low-Dimens. Syst. Nanostruct. **68**, 33 (2015)
9. Lin, Y.-M., et al.: Science **327**, 662 (2010)
10. Lee, J., et al.: ACS Nano **7**(9), 7744–7750 (2013)
11. Yan, F., Zhang, M., Li, J.: Adv. Healthc. Mater. **3**, 313 (2014)
12. Bergveld, P.: Development of an ion-sensitive solid-state device for neurophysiological measurements. IEEE Trans. Biom. Eng. **MBE-17**, 70 (1970)
13. Fang, T., Konar, A., Xing, H., Jena, D.: Appl. Phys. Lett. **91**, 092109 (2007)
14. Giannazzo, F., Sonde, S., Raineri, V., Rimini, E.: Nano Lett. **9**, 23 (2009)

15. Zhu, Y., Wang, C., Petrone, N., Yu, J., Nuckolls, C., Hone, J., Lin, Q.: 2015 28th IEEE International Conference Micro Electro Mechanical Systems 2015-Feb, vol. 869 (2015)
16. Ebrish, M.A., Shao, H., Koester, S.J.: Appl. Phys. Lett. **100**, 143102 (2012)
17. Koester, S.J.: Appl. Phys. A **99**, 163105 (2011)
18. Giannazzo, F., Sonde, S., Lo, R., Nigro, E., Rimini, V.: Nano Lett. **11**, 4612–4618 (2011)

Silicene in the Flatland

C. Grazianetti and A. Molle

Abstract Since the rise of graphene, the flatland of two-dimensional (2D) materials continues to expand its borders including more and more members with complementary properties. An overview of the more relevant members is proposed which accounts for the class of 2D layered transition metal dichalcogenides and elementary 2D materials. Amidst the latter ones is silicene, a honeycomb-like Si lattice. Silicene has recently attracted an enormous interest as emerging research material for the semiconductor technology roadmap because of its intrinsic affinity with the ubiquitous silicon technology. Free-standing silicene is energetically allowed provided that the chemical bonds are vertically buckled rather than being perfectly planar as in graphene. Nonetheless, a stable graphite-like Si allotrope does not exist in nature. Artificially forcing silicon atoms into a silicene lattice is made possible by the epitaxy of a silicon monolayer on substrates therein giving rise to a quite rich variety of surface phases. Here we review the phase diagram of silicene on Ag(111) as a paradigmatic case of the silicene-on-substrate. Attention is also paid to identify templates for the silicene growth enabling the silicene transfer to technological platforms targeting device integration.

1 An Overview on 2D Materials for Nanoelectronics

The debut of graphene [1] paved the way to the exploration of an "expanding universe" of two-dimensional (2D) materials that may serve as building blocks for a multifunctional nanotechnology or as a playground for an undiscovered physics. The gate modulation and the exceptionally high carrier mobility in graphene transistors prompted an enormous interest in the integration of 2D materials into ultimately scaled nanoelectronic devices [2, 3]. In this framework, graphene, i.e., a 2D lattice of carbon atoms arranged in hexagonal rings and kept together by

C. Grazianetti · A. Molle (✉)
CNR-IMM, Unit of Agrate Brianza, via C. Olivetti 2, 20864 Agrate Brianza, MB, Italy
e-mail: alessandro.molle@mdm.imm.cnr.it

V. Morandi and L. Ottaviano (eds.), *GraphITA*, Carbon Nanostructures,
DOI 10.1007/978-3-319-58134-7_10

in-plane sp^2 hybridized bonds, has a limited range of applications as long as it behaves like a zero-gap semiconductor with the conduction and valence bands having a characteristic linear dispersion in the proximity of the K points of the hexagonal Brillouin zone. If integrated as an active channel in a conventionally planar transistor architecture, graphene looks promising for radiofrequency (RF) applications, but offers poor performances for ultra-fast digital devices because of the absence of a sizeable energy gap allowing for the sharp commutation of the logic state [4]. Other 2D materials have come into play, which may overcome the intrinsic limitations of graphene for each application where a fast logic switch is demanded [2]. Figure 1 illustrates an overview of 2D materials beyond graphene that can fit to that scope. Thus far the more promising option in this respect relied on single or few layers of MoS_2 [5]. In its more stable form, MoS_2 is a layered transition metal dichalchogenide (TMD) with a trigonal prismatic (2H) coordination of the Mo atoms, where the MoS_2 layers are internally stacked together by weak interplane van der Waals (vdW) forces, and atoms in each individual MoS_2 layer are covalently bonded [6]. 2H-MoS_2 is a semiconductor with a direct gap (1.80 eV) when it is reduced to a single layer, or with an indirect gap (1.29 eV) in its multilayer form. A field effect transistor (FET) based on a MoS_2 single layer was pioneered on 2011 therein proving an electron mobility up to 200 cm^2/Vs, an ON/OFF ratio of 1×10^8 in its transfer characteristic, and a very steep

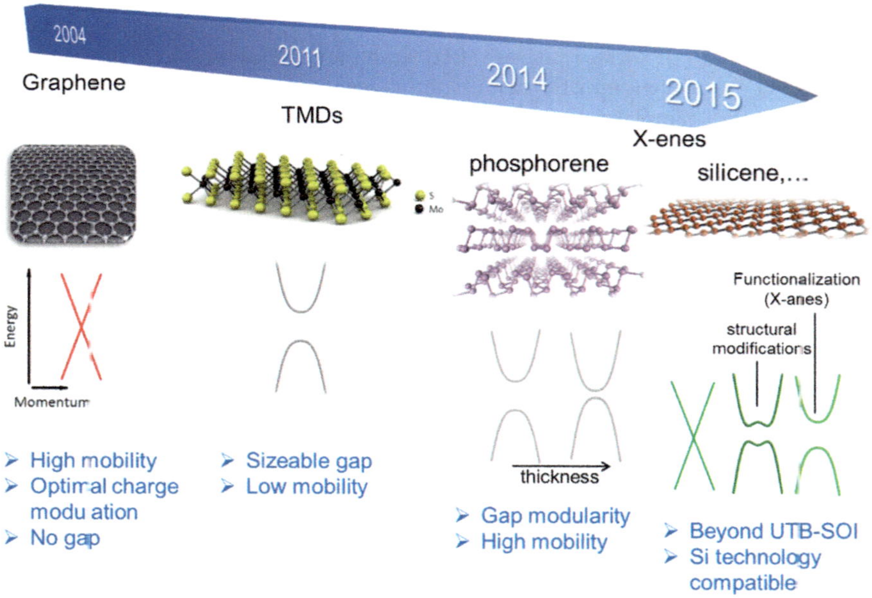

Fig. 1 Overview of the 2D materials suitable for nanoelectronic applications. Each considered materials is presented with a sketch of its characteristic structure, energy band structure, and the salient properties for electronic devices are reported. The main arrow on top draws the chronological sequence of each material integration into an operating device

subthreshold swing at room temperature that makes it suitable for ultra-low power applications [7]. The intrinsic presence of an optical gap in the MoS_2 single layer also enabled it to access optoelectronic applications, such as phototransistors or photodiodes [8, 9]. However, massive integration of MoS_2 in nanoelectronics is currently jeopardized by the difficulty of synthesizing MoS_2 nanosheets on a macroscopic (e.g., wafer-scaled) area. Originally, MoS_2 nanosheets were extracted by means of mechanical exfoliation on the microscale. This methodology proved to be straightforward for the investigation of fundamental physics or for the verification of prototypical devices, but exfoliated flakes of natural MoS_2 crystals are clearly inadequate for a large-scale production and prone to native contamination [10, 11]. Nowadays ongoing efforts are carried out to develop simple and cost-effective chemical vapor deposition methods yielding wafer-scaled MoS_2 with controlled number of layers [12–14]. To a more general overview, MoS_2 is a member of the wide family of layered TMDs that combine (group IV, V, VI, up to X) transition metals with chalcogens so as to give a variety of electronic structures including the semiconducting, metallic, semimetallic, charge density wave or superconducting states depending on the filling of their more external *d* shell [6]. Among them, WSe_2 was recently shown to provide the higher mobility for high-speed logic applications [15]. The variability of the electronic configuration throughout the TMD family also allows for engineering vdW heterostructures made of different TMD constituents with tailored band line-up so as to create 2D interface regions suitable for tunnel FET operations [16].

An emerging competitor of MoS_2 in nanoelectronic applications is phosphorene, a term usually adopted to denote single or few layers of black phosphorous (BP) [17]. BP is a stable allotrope of phosphorous originally obtained by exposing white phosphorous at high pressure and temperature. Similar to graphite, the BP layered structure implies interlayer vdW forces and strong intralayer covalent bonding, but unlike graphene, phosphorene exhibits a puckered honeycomb lattice structure with an in-plane anisotropy. From the electronic point of view, the more salient feature of phosphorene is to offer an optical gap modularity as a function of the number of layer from 1 eV in a single layer at the Γ point down to 0.3 eV in multilayer at the Z point. On the other hand, a highly anisotropic electrical transport was reported on phosphorene with a record field effect mobility amounting to $\sim$1000 cm^2/Vs through a 10 nm-thick BP channel at room temperature. The relatively high mobility and the gap tunability makes phosphorene a versatile material of interest for multiple applications including electronics, optoelectronics (especially in the infrared regime), and photovoltaics [17–20]. Despite these advantages and in analogy with MoS_2, the synthesis of BP is currently confined to exfoliation methods that severely limit the production to micro-scaled flakes far away from a large-scale fashion.

As much as graphene, phosphorene can be referred to as belonging to the general class of X-enes materials, namely elementary 2D materials with a single X constituent and a hexagonal lattice symmetry [21]. In reality, the term X-ene has recently come up to the attention in order to denote artificially grown 2D elementary lattices [22]. These materials are usually known as silicene, germanene, and stanene, i.e., the graphene alias of Si, Ge, and Sn. Recently 2D metallic boron adds to this list [23, 24]. However, unlike graphene, X-enes are not stable in nature as due to their tendency to form sp^3 hybrid bonding instead of sp^2 ones, but they can be epitaxially grown on substrates. The toll to pay is the emergence of a vertically buckled structure deviating from the perfect planarity of graphene. The first X-ene to be integrated into a FET device was silicene, therein proving an ambipolar behavior with a carrier mobility up to 100 cm^2/Vs at room temperature [25]. Despite the relatively low mobility compared to graphene, the proof of a silicene transistor outlined the opportunity to go beyond the limit of the complementary metal oxide semiconductor (CMOS) technology based on ultra-thin silicon film channels bringing silicon into electronic device at the true 2D limit and opening new pathways for the implementation of other X-enes in a nanoelectronic framework. According to the theoretical predictions, the ambipolar behavior of the silicene transistor is apparently reminiscent of a graphene-like semimetallic character and would suffer from the same limitation of graphene. However, buckling in silicene as well as chemical functionalization of silicene can be used to induce a band gap opening that would eventually make it appealing for logic applications [26].

2 Silicene and the Class of X-enes

The effort to reduce silicon to the 2D limit of a silicene lattice is not merely a matter of scientific curiosity but it is also driven by technology issues. Given its affinity with silicon and its potential compatibility with the ubiquitous silicon-based technology, silicene has attracted a tremendous consideration since its original first-principle prediction as a 2D buckled lattice dating back to Takeda et al. [27] and then revisited by Cahangirov et al. [28]. From these early models, the concept of silicene encountered the requirement of an ultimate dimensionality reduction of the channel size feature in a CMOS logic because it would help bringing the conventional Si-based MOSFET to the extreme miniaturization [29]. The continuous lateral scaling of Si nanoelectronic devices and circuits demands for a dramatic reduction of the active channel thickness in order to improve the device electrostatics. Therefore, silicon with a reduced dimensionality, herein including silicon nanowires as well as 2D silicon, are potentially regarded as new emerging device materials targeting the next nodes of the semiconductor technology roadmap [30]. This scenario is depicted in Fig. 2 as a technology evolution of the silicon-based devices towards 2D materials as future solutions. Ultra-thin silicon channel devices including the ultra-thin body (UTB) silicon-on-insulator (SOI) channels in a conventional FET architecture or the gate surrounded vertical channels in FinFET are the current options for the ultrascaled CMOS technology. New and perhaps revolutionary solutions, either in terms of emerging research materials or devices, are needed to face further scaling requirements (according to an acceleration of the well-known Moore's law, known as "More Moore"

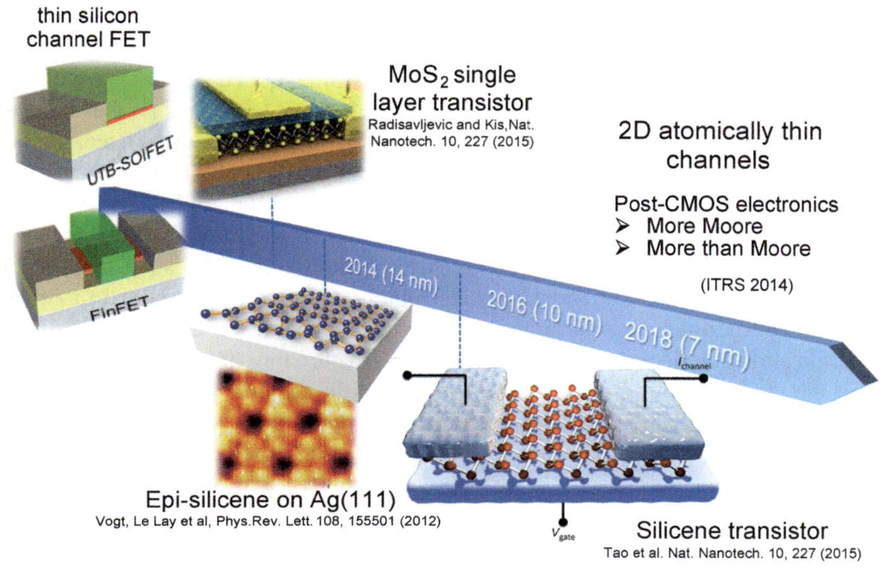

Fig. 2 Technology evolution of logic devices incorporating ultra-thin channels towards the potential integration of 2D atomically thin channels (the main arrow marks the technology nodes according to the ITRS 2014) [30]. Starting from the current mainstream device architectures (UTB-SOIFET and FinFET), a turning point on this roadmap dates back to 2011 when a MoS_2 single layer was successfully integrated into a FET by Radisavljevic et al. [7] thus paving the way to 2D materials for nanoelectronics. A subsequent scientific breakthrough was the first compelling evidence of the epitaxial silicene on Ag(111) substrates reported by Vogt et al. on 2012 [32]. Silicene made its debut in a transistor on 2015 as reported by Tao et al. [25]. The reduced dimensionality in silicon channels is part of roadmap of 2D atomically thin channels for nanoelectronics where Post-CMOS requirements such as the "More Moore" and the "More than Moore" issues are addressed [30]

approach) and the demand for multiple functionalities on a single device (a trend known as "More than Moore"). In this respect, the 2D options mentioned in Fig. 1 constitute an interesting materials platform where to boost performance of a planar CMOS technology or to define different device architectures, such as the tunnel FET [31]. In particular, a research breakthrough in the silicon dimensionality reduction was marked by the proof of the epitaxial silicene on Ag(111) template as reported by Vogt et al. on 2012 [32]. Despite the disruptive evidence of a stabilized silicene layer on a substrate, the debut of a silicene transistor came only in 2015 after developing a protocol where the silicene monolayer is peeled off from its native Ag substrate and flipped onto a device-friendly SiO_2/Si substrate therein enabling the gate bias control on the silicene channel [25]. In the meanwhile, other X-enes or functionalized X-enes appeared on the stage as a result of a physical growth on appropriate substrate (in substantial agreement with the silicene epitaxy) [22] or via chemical methods [33].

Along the line of graphene, the existence of silicene and other X-enes pose additional fundamental questions that may open new avenues in the future advances of condensed matter physics. Apart from the above-mentioned technology impulse, not only the existence of the X-enes seem like a "freak of nature" as long as none of the X-ene beyond graphene is existing in nature as free-standing material, but X-enes are also envisioned to be a carrier of new exotic phenomena such as the quantum spin Hall effect or other topology-related phenomena [21, 34]. Although X-enes are expected to share the hexagonal honeycomb structure of graphene, some significant differences can be singled out while spanning the elements from C to Sn. The hybridization of the chemical bonds is purely sp^2 in graphene, mixed sp^2/sp^3 in silicene and germanene, and almost fully sp^3 in stanene [28, 32, 35]. This mixed hybridization is reflected on the vertically buckled structure of free-standing X-enes (see the sketch in Fig. 3a) as results from ab initio calculations [28]. Despite the intrinsically buckled lattice, the calculated band structure of X-enes generally results in linearly dispersed bands at the K/K' points of the Brillouin zone hosting Dirac fermions in close resemblance with graphene. As such, X-enes do not seem to bring significant advantages compared to their carbon "forerunner." However,

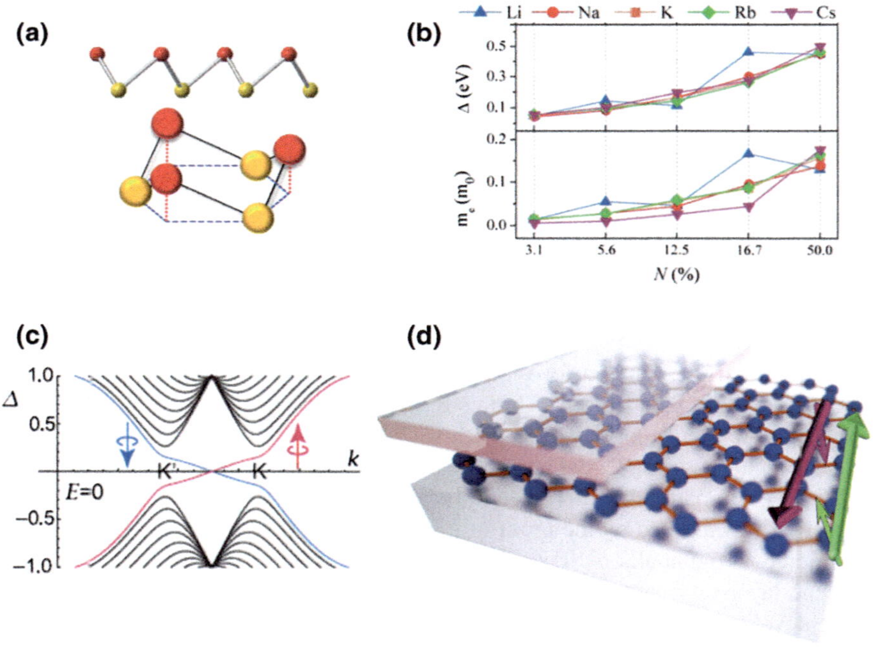

Fig. 3 **a** Buckled silicene lattice: side view and lateral view. **b** Band gap engineering in silicene via incorporation of foreign atoms. Adapted from Ref. [36]. **c** Band structure model of a silicene at the K and K' points where a topological gap opens and helical edge states set inside the gap. Adapted from Ref. [34]. **d** Pictorial sketch of an X-ene lattice where helical and non-dissipative conduction occurs at the edge

the stronger reactivity stemming from the mixed hybridization is expected to result in significant modification of the X-enes band structure upon incorporation of foreign atomic species. For instance, silicene was predicted to undergo a substantial band gap opening when hosting alkali metals in its structure o after chemisorption with oxygen or hydrogen [36, 37].

An additional feature of X-enes comes out from their nontrivial topology. This aspect descend from the Kane-Mele model describing graphene as an intrinsic 2D topological insulator because of the concomitant opening of topological energy gap due to the spin-orbit coupling term in the graphene Hamiltonian and the emergence of edge states linearly dispersed in the topological gap (see the diagram in Fig. 3c) [38]. 2D topological insulators identify a state of matter where the spin and the momentum in the a 2D lattice are mutually locked so as to give an inner surface with an insulating character surrounded by metallic borders (see the representative picture where the edge of a generic honeycomb X-ene lattice bears the spin-polarized 1D conduction in Fig. 3d) [39]. Moreover, moving from C to Sn implies an increase of the atomic mass and a consequent strengthening of the spin-orbit coupling that leads to a larger energy gap opening at the Dirac point up to an expected value of 0.1 eV in stanene [40, 41]. The recent finding of a stanene lattice grown on a Bi_2Te_3 substrate [42] prompted a renewed interest on the realization of a 2D topological insulator whose signature was so far detected only on HgTe/CdTe quantum well structures at temperature below 10 K [43].

3 A Case in Point: The Phase Diagram of the Epitaxial Silicene

Among the many substrates used for epitaxial growth of silicene [22, 44], most of the experimental and theoretical work on the silicene epitaxy on substrates has been carried out on the Ag(111) template, as natural continuation of the growth of silicene nanoribbons on Ag(110) [22, 41, 45]. Moreover, Ag(111) satisfies the requirements of being commensurate with the predicted stable free-standing silicene showing sixfold top-layer symmetry in order to host a honeycomb Si lattice and showing a moderate chemical reactivity with Si adatoms [32]. A *posteriori*, it can be demonstrated that the requirements on the lattice match can be slightly relaxed. In fact, at variance with graphene, epitaxial silicene is not planar. Therefore, the incoming Si atoms self-organize in structurally different superstructures, i.e., different atomic arrangements. This intrinsic flexibility is related to the tendency of silicene to pucker in vertically buckled structures [28]. Hence, the buckling acts as an additional degree of freedom. This specific feature can then be referred to as the *polymorphic nature* of silicene, i.e., the skill of Si atoms to adapt to the substrates in several buckled structures with different periodicities. In particular, here, we refer to the polymorphic nature of silicene because different silicene superstructures may coexist on the same Ag substrate [22]. Moreover, we will also refer to these structures as silicene phases because they can be rationalized in terms of a thermodynamic surface phase diagram as discussed in the following. Silicene on Ag (111) gives rise to four main phases, whose structures depend on the commensurability relation with the substrate surface and on the allowed buckling distributions. A picture of four representative silicene phases is displayed in Fig. 4 by combining hard-spheres models with the corresponding high-resolution scanning tunneling microscopy (STM) topographies.

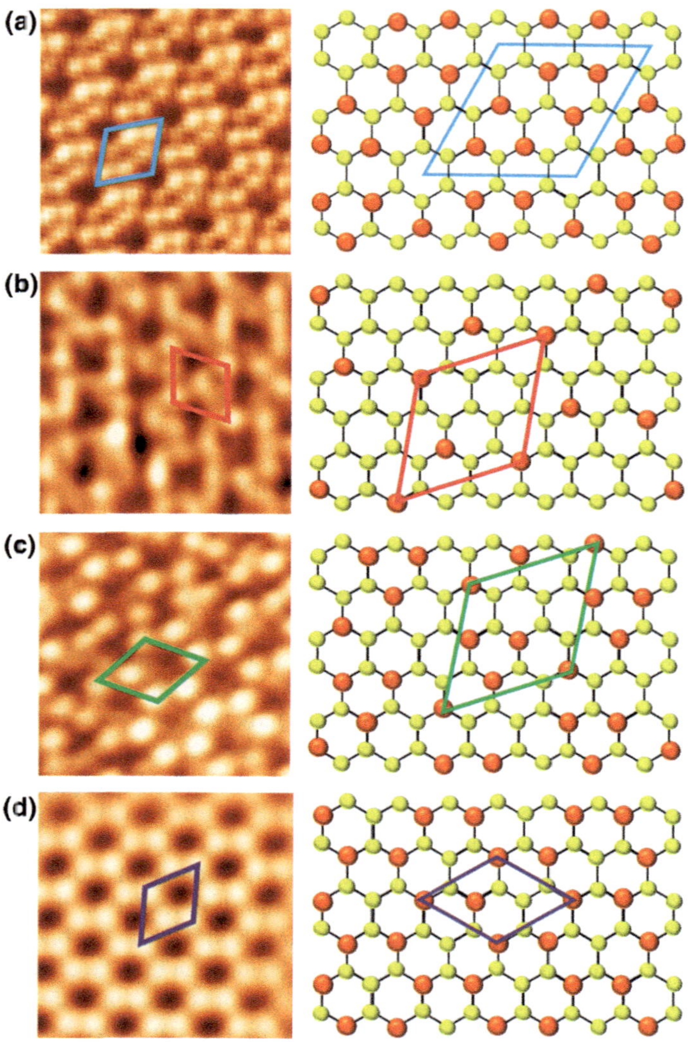

Fig. 4 Silicene phases on Ag(111). **a** 4×4, **b** $\sqrt{13} \times \sqrt{13}$, **c** $2\sqrt{3} \times 2\sqrt{3}$, and **d** $\sqrt{3} \times \sqrt{3}$ phases. *Left column* show real STM images while *right column* depicts the respective hard-spheres models. Rhombi indicates unit cells

Figure 4a illustrates the very first silicene phase ever observed on Ag(111), termed as the 3 × 3 silicene on 4 × 4 Ag (hereafter 4 × 4 silicene in order to avoid cumbersome notation) [32]. By inspecting the atomically resolved STM image, it is hard to recognize a honeycomb-like pattern, because of the higher STM sensitivity to record the buckled atoms as protrusions. Nonetheless, the identification of a hexagonal and buckled structure is made possible from symmetry considerations and by matching it with the hard-spheres model. The so-called flower-like pattern, characterized by triangular structures hexagonally arranged around the dark holes, is given by the Si atoms located either on top or in-between the Ag atoms. Hence, the bright spots in the STM images do not represent every single atom in silicene. Dark holes imaged by STM are separated by an average value of 1.14 nm, i.e., four times the surface Ag(111) lattice constant (0.289 nm) [32, 46]. The 4 × 4 silicene buckling is quite large (0.71–0.79 Å) and asymmetric, [47] thus significantly deviating from the relatively low vertical distortion of the free-standing silicene as modelled in Ref. [28]. In contrast with graphene which is purely sp^2 hybridized, the Si–Si bonding character of silicene is mixed sp^2/sp^3. Indeed, for the 4 × 4 case, ab initio calculations determined that bond angles are 110° for the six top Si atoms (very close to an angle of 109.5° for an ideally sp^3-hybridized Si atom), 120° for the six bottom Si atoms (purely sp^2 hybridized) and the other six bottom Si atoms have bond angles of between 112° and 118° [32]. The atomic structure model of $\sqrt{13} \times \sqrt{13}$ phase is illustrated in Fig. 4b. The $\sqrt{7} \times \sqrt{7}$ silicene on the $\sqrt{13} \times \sqrt{13}$ Ag(111) surface (hereafter $\sqrt{13} \times \sqrt{13}$ silicene only) is a supercell of silicene made up of 14 atoms and rotated by approximately 13.9° with respect to the $[110]_{Ag}$ direction [22]. In contrast with the 4 × 4 silicene, the $\sqrt{13} \times \sqrt{13}$ silicene is regularly buckled and its buckling is of 0.79 Å very close to that of the 4 × 4 superstructure [39]. The $2\sqrt{3} \times 2\sqrt{3}$ superstructure ($\sqrt{7} \times \sqrt{7}$ silicene reconstruction) with a periodicity of 1 nm is the third of the commonly observed phases in a silicene monolayer (Fig. 4c). This phase is characterized by a buckling of 1 Å and it is misaligned from the 4 × 4 phase by an angle of 30° [43, 44]. Although this phase has been observed to coexist with both 4 × 4 and $\sqrt{13} \times \sqrt{13}$ [48], typically it can be found with large domains when Si is deposited onto the Ag(111) surface at higher temperatures [44, 49]. While the previous phases are observed in monolayer silicene only, there is an additional phase, $\sqrt{3} \times \sqrt{3}$ silicene reconstruction, namely a $4/\sqrt{3} \times 4/\sqrt{3}$ Ag superstructure, which was reported as either monolayer or multilayer silicene (see Fig. 4d) [50]. Moreover, in the multilayer silicene case, any layer over the first one reproduces a $\sqrt{3} \times \sqrt{3}$ reconstruction up to tens of monolayers (ML). The $\sqrt{3} \times \sqrt{3}$ phase shows a periodicity of 0.64 nm and the separation between the two consecutive $\sqrt{3} \times \sqrt{3}$ layers is 0.3 nm [47]. The evidence of multiple silicene phases on the same substrate can be sorted in a 2D phase diagram as a function of the main thermodynamic quantities during growth, the chemical potential—dictated by the number of Si adatoms extrinsically supplied to the Ag(111) surface (namely, the coverage)—and the substrate temperature.

Figure 5 summarizes the silicene phase diagram as function of coverage (in ML) and substrate temperature. In the sub-monolayer coverage regime (<1 ML),

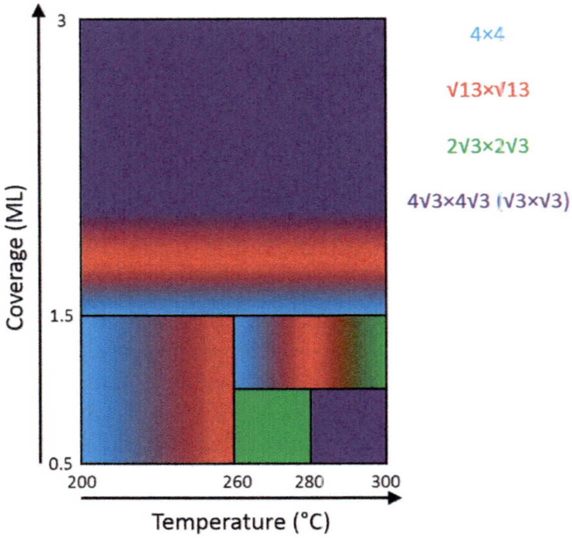

Fig. 5 Phase diagram of silicene phases on Ag(111) as function of coverage and substrate temperature

a strong dependence on the substrate temperature is found. In the temperature range comprised between 200 and 260 °C, there is coexistence of 4 × 4 and $\sqrt{13} \times \sqrt{13}$ phases. It should be pointed out that these two phases are always found to be coexisting inside the phase diagram of Fig. 5. Largely dominant phases such as the $2\sqrt{3} \times 2\sqrt{3}$ and the $\sqrt{3} \times \sqrt{3}$ phases can be singled out by increasing the substrate temperature up to 280 and 300 °C, respectively. Although the phase diagram is supported by an extensive STM analysis, we cannot rule out the presence of minor traces of other silicene phases and, therefore, we refer to the dominant phase as a statistical concept. In the coverage regime between 1 and 1.5 ML, the phase diagram is characterized by the threshold temperature of 260 °C. Below such a temperature, 4 × 4 and $\sqrt{13} \times \sqrt{13}$ phases appear in close analogy to sub-monolayer case, whereas above there is coexistence of all the three silicene monolayer phases. In this thickness regime and in the whole temperature range considered, small (few nm) second layer islands made of $\sqrt{3} \times \sqrt{3}$ start growing. For coverage above 1.5 ML the phase diagram does not depend on the substrate temperature any longer. Indeed, in the considered substrate temperature regime (200–300 °C), silicene growth is characterized by 4 × 4 and $\sqrt{13} \times \sqrt{13}$ phases as seeding layer, over which $\sqrt{3} \times \sqrt{3}$ reconstructed layers pile up through island formation. Although the exact growth mode, either Stransky–Krastanov or Volmer–Weber, has not been clearly determined, the condition for multilayer silicene growth is determined by the presence of a silicene monolayer as a seed layer.

In addition to the reported phase diagram, the variety of the silicene phases can be tuned by means of post-deposition annealing (PDA) processes [51]. Temperature-driven transitions occur in a narrow temperature range, thus demonstrating the metastable character of silicene, which is governed by a kinetically limited growth [48, 51]. As a starting point, consider a silicene sample grown at

250 °C in the monolayer coverage regime, where there is coexistence of 4×4 and $\sqrt{13} \times \sqrt{13}$ phases. After PDA at 270 °C for 30 min and quenching at room temperature, large domains of $2\sqrt{3} \times 2\sqrt{3}$ phase can be found. At this stage, a phase transition takes place, which involves a re-arrangement of the silicene periodicity. It is important to stress that the PDA process started from room temperature to a higher temperature (270 °C) with respect to the growth one (250 °C), thus confirming the (meta)stable character of the new silicene phase obtained after cooling down to room temperature. A second PDA at higher temperature (290 °C) results in a further phase transition characterized by the presence of large domains of $\sqrt{13} \times \sqrt{13}$ superstructure. In the considered thickness regime, i.e., 1ML coverage, the $\sqrt{13} \times \sqrt{13}$ phase appears again only after the second PDA as a dominant phase whereas it was usually observed to coexist with the 4×4 in the as-grown silicene layer. For a growth temperature of 250 °C and nominal thickness of 1 ML, the starting phase is mainly featured by an interplay of the 4×4 and $\sqrt{13} \times \sqrt{13}$ phases which, after a first PDA at 270 °C, transform into a $2\sqrt{3} \times 2\sqrt{3}$ phase. After a second PDA at 290 °C, the dominant phase becomes the $\sqrt{13} \times \sqrt{13}$ again. Eventually, a different phase is recovered when the PDA temperature is directly increased up to 290 °C without passing through the intermediate stage at 270 °C. In this case, the observed phase is the $4/\sqrt{3} \times 4/\sqrt{3}$ phase. This phase transition is typically related to dewetting mechanisms [51]. It is possible to observe that the described phase transitions follows a trend, which is different from that reported in the phase diagram of Fig. 5 for the as-grown phases. As aforementioned, the PDA tuning of the silicene phase is potentially interesting for band gap engineering as far as different silicene phase are expected to have different electronic structure when disentangled from the supporting substrate [26].

4 Towards Device Integration of Silicene and Silicene-like Materials

Integration of a silicene monolayer into a device structure is basically hurdled by the intrinsic coupling of silicene with what we may denote as its top and bottom interfaces. The top interface is the exposed silicene surface when it is epitaxially grown on a substrate. In such a case, the top silicene interface is prone to easy degradation and eventually to oxidation if exposed to environmental condition. Even on a defect-free silicene lattice, the fast degradation is intrinsically caused by the mixed sp^2/sp^3 hybridization which makes silicene be chemically reactive. As such, no solution is possible but for a protective encapsulation of the silicene enough to avoid any ambient reaction. In order to prevent oxidation, encapsulation of the Ag-supported silicene was made possible by capping the silicene with an ultra-thin (few nm) Al_2O_3 layer grown in situ, namely in the same ambient of the silicene epitaxy [52]. A sketch of the Al_2O_3 encapsulation is illustrated in Fig. 6a for a silicene on a substrate (the kind of substrate will be next matter of discussion).

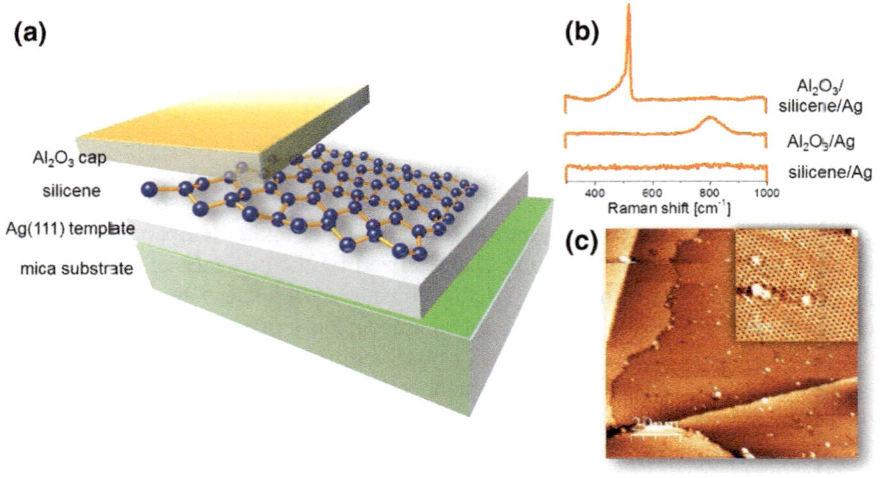

Fig. 6 **a** Pictorial sketch of an encapsulated silicene-on-substrate structure where the encapsulation consists of an ultra-thin Al_2O_3 capping layer, and the Ag-based substrate is a cleavable epi-Ag on mica. **b** Comparative Raman spectroscopy study that highlights the permanence of silicene in the Al_2O_3/silicene/Ag structure. **c** STM image of the atomic structure of silicene grown epi-Ag/mica susbtrate. *Inset* magnification of the 4 × 4 silicene phase

The effectiveness of the Al_2O_3 capping layer is demonstrated by the ex situ Raman spectroscopy study in Fig. 6b where an encapsulated silicene sample is compared with silicene-free and an uncapped silicene samples. In the former case, the observation of sharp Raman peak was interpreted as the hallmark of a buckled silicene lattice thus making evidence of the silicene survival upon encapsulation [44, 47].

On the bottom side, the main issue relies on how to strip the silicene off from its Ag(111) substrate. From practical point of view, the Ag(111) monocrystal is undesired because it is extremely expensive and inadequate for any device-oriented exploitation of silicene. Although Ag works as good template for the silicene growth, an additional, but not less important, reason to detach it from silicene comes from the strong hybridization between Si and Ag atoms that results in a net metallization of the silicene and in the suppression of the Dirac cones [53]. One way to address the substrate issue is to use a modified Ag substrate where to more comfortably handle the silicene monolayer after encapsulation. The introduction of cleavable 300 nm-thick epi-Ag(111)/mica substrates for the silicene epitaxy has definitely relaxed this hurdle thereby marking a sharp technological breakthrough in the silicene handling. In this respect, it must be emphasized that that silicene grown on Ag(111)/mica replicates the same features as when it is grown on Ag(111) monocrystal substrates. This fact is reflected in the atomically resolved STM image in Fig. 6c making evidence of the formation of an extended 4 × 4 phase on top of the terraces of a commercially provided 300 nm-thick epi-Ag(111) film on a mica substrate (see the inset for the magnification of the 4 × 4 silicene structure).

The interplay of the Al_2O_3 encapsulation with the implementation of cleavable Ag/mica is at the bottom of the so-called silicene encapsulated delamination with native electrodes (SEDNE) process proposed by Tao et al. to fabricate and operate a silicene transistor [25]. This methodology is illustrated in Fig. 7a in its sequential steps, where the final step corresponds to a bottom gate FET based on a silicene channel. Specifically, the SEDNE process flow starts with the encapsulated silicene being delaminated from its pristine substrate therein resulting in a sheet where the silicene monolayer is sandwiched in-between the Al_2O_3 capping layer and the residual Ag support. The "silicene sandwich" is then placed upside down on a

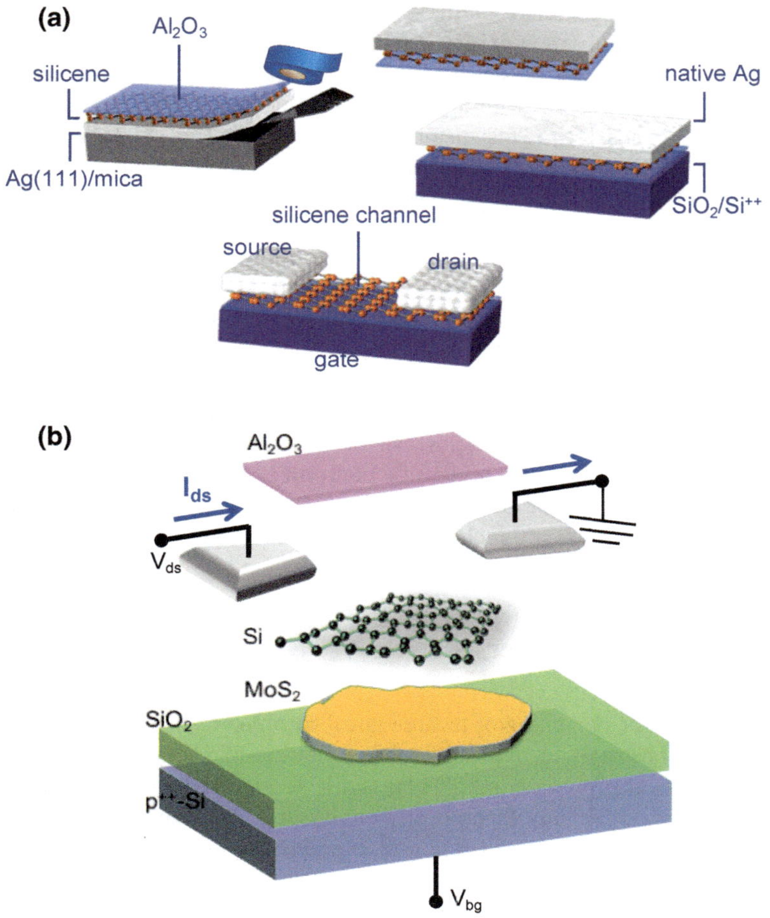

Fig. 7 Silicene and silicene-like-based transistor. **a** Schematic process steps of silicene on Ag (111) and **b** 2D Si on MoS_2 substrate. In both devices architectures encapsulation is achieved by room temperature deposition of Al_2O_3. Panels **a** and **b** are adapted from Refs. [25] and [56], respectively

"device-friendly" SiO_2/Si substrate (bottom gate dielectric). Source and drain electrodes are then lithographically designed on the native Ag overlying layer and the residual Ag outside the patterned region is chemically etched off resulting in an Ag-free silicene conductive channel.

Substrate engineering for the silicene epitaxy is an alternative way to address the substrate issue. This approach is particularly interesting for those non-metallic substrates that act as templates for the silicene growth without strongly interacting with silicene itself as metallic substrate like Ag does. To this purpose, other 2D materials such hex-BN or TMDs have been proposed as for a vdW epitaxy of silicene [54], namely an epitaxial growth where the silicene adlayer would weakly interact with the underlying substrate via vdW forces. An example of this approach is the growth of silicene-like 2D nanosheets on MoS_2 substrate [55]. Despite the weak interaction, silicene is observed to significantly pucker throughout the MoS_2 surface. Nonetheless, the presence of a silicene nanosheet has a strong influence in the electronic band line-up with the MoS_2 substrate therein resulting in a distortion of the MoS_2 bands [56]. This distortion has been used to create a 2D Si/MoS_2 heterosheet interface channel to integrate in a FET device structure. In detail, the Si/MoS_2 heterosheet FET was based on silicene nanosheet grown onto a mechanically exfoliated MoS_2 multilayer flakes onto SiO_2/Si substrates (utilized as a bottom electrode) [56]. In close analogy with silicene on Ag(111), encapsulation with a few nm-thick Al_2O_3 was made by reactive MBD. In particular, the electrical response of such a transistor is characterized by an active channel at the Si/MoS_2 interface with an effective n-type character, which confirms the electron confinement detected by photoelectron spectroscopy experiments [56].

The two cases reported in Fig. 7 identify a similar platform on which different silicene-based device configurations can be fabricated, despite the different substrate used. In particular, such a versatile platform is devised on the Al_2O_3 capping layer which successfully protects Si films against oxidation and serves as dielectric gate without impacting on the chemical and electronic properties of Si channels.

5 Conclusions and Perspectives

We may define the beginning of technological roadmap of X-enes from 2012 when Vogt et al. pioneeringly recognized a subtle graphene-like honeycomb structure in a silicon monolayer grown on a Ag(111), and hence revived a concept of a silicene lattice that was so far only theoretically envisioned [32]. Since that time, silicene has received a booming consideration in experimental and theoretical works. The strong hybridization that the epitaxial silicene proved to experience with almost all its hosting metallic substrates appeared as a critical bottleneck for any practical application. This drawback would have progressively confined silicene to an academic consideration without any real impact, but the recent integration into a device structure has scored a breakthrough for a possible silicene technology. On the other hand, silicene cousins such as germanene and stanene, appeared on the stage thus paving the way to the potential realization of the elusive 2D topological insulator state. Restricting the attention to silicene only, a big deal of work is still left in terms of material optimization and validation. The silicene transistor showed room temperature operation but the measured mobility turned out to be significantly lower than that theoretically expected from an ideal silicene layer. The reason may come from the concentration of point defects or line defects that inherently affect the silicene structure. Further optimization are then demanded to assess the superior limit that can be reached by the carrier mobility through a silicene channel. On the other hand, the silicene transfer methodology also needs further improvements to consolidate a technologically reliable integration process into a device structure. Ongoing efforts in this respect are currently addressing the synthesis of silicene via chemical methods or the substrate engineering for the silicene epitaxy. This scenario depicts the state of the art in the silicene as an incubation stage where preliminary results are tremendously encouraging to envision a silicon technology at the 2D limit, but no definite outcomes can still be drawn. Overall, despite the latest effort to replace the "old" silicon with other mobility boosters such as germanium or III-V semiconductors, the advent of silicene appears to be a kind of renaissance of the evergreen silicon still enabling a dimensionally scaled evolution of the planar CMOS technology of digital devices.

Acknowledgements The Authors acknowledge support from the CNR grant "Joint Lab": http://www.cnr.it/sitocnr/Englishversion/CNR/Activities/IntActivity/JointLaboratory.html.

References

1. Novoselov, K., et al.: Science **306**, 666 (2004)
2. Fiori, G., et al.: Nat. Nanotech. **9**, 768 (2014)
3. Houssa, M., Dimoulas, A., Molle, A. (eds.): 2D Materials for Nanoelectronics. CRC Press (Taylor & Francis Group), 2016 Boca Raton FL
4. Schwierz, F.: Nat. Nanotech. **5**, 487 (2010)
5. Jariwala, D., Sangwan, V.K., Lauhon, L.J., Marks, T.J., Hersam, M.C.: ACS Nano **8**, 1102 (2014)
6. Chhowalla, M., et al.: Nat. Chem. **5**, 263 (2013)
7. Radisavljevic, B., et al.: Nat. Nanotech. **6**, 147 (2011)
8. Zin, Y., et al.: ACS Nano **1**, 74 (2012)
9. Kufer, D., Konstantatos, G.: Nano Lett. **15**, 7307 (2015)
10. Molle, A., et al.: Adv. Electr. Mater. **2**, 1600091 (2016)
11. Addou, R., et al.: ACS Appl. Mater. Interf. **7**, 11921 (2011)
12. Kang, K., et al.: Nature **520**, 656 (2015)
13. Najmaei, S., et al.: Nat. Mater. **12**, 754 (2013)
14. Vangelista, S., et al.: Nanotechnology **27**, 175703 (2016)
15. Chuang, H.J., et al.: Nano Lett. **14**, 3594 (2014)
16. Gong, C., et al.: Appl. Phys. Lett. **103**, 053513 (2013)
17. Li, L., et al.: Nat. Nanotech. **9**, 372 (2014)
18. Liu, H., et al.: ACS Nano **8**, 4033 (2014)
19. Zu, W., et al.: Nano Lett. **15**, 1883 (2015)
20. Viti, B., et al.: Adv. Mater. **27**, 5567 (2015)

21. Molle, A., et al.: Nat. Mater. **16**, 163 (2017)
22. Grazianetti, C., et al.: 2D Mater. **3**, 012001 (2016)
23. Mannix, A.J., et al.: Science **350**, 1513 (2016)
24. Feng, B., et al., Nat. Chem. **8**, 563 (2016)
25. Tao, L., et al.: Nat. Nanotech. **10**, 227 (2015)
26. Scalise, E., et al.: Appl. Surf. Sci. **291**, 113 (2014)
27. Takeda, K., Shiraishi, K.: Phys. Rev. B **50**, 14916 (1994)
28. Cahangirov, S., et al.: Phys. Rev. Lett. **102**, 236804 (2009)
29. Ahmed, K., Schuegraf, K.: See the article: 'Transistor Wars'. http://spectrum.ieee.org/semiconductors/devices/transistor-wars
30. International Technology Roadmap for Semiconductors, see: http://www.itrs2.net/itrs-reports.html
31. Sarkar, D., et al.: Nature **526**, 91 (2015)
32. Vogt, P., et al.: Phys. Rev. Lett. **108**, 155501 (2012)
33. Bianco, E., et al.: ACS Nano **7**, 4414 (2013)
34. Ezawa, M.: New J. Phys. **14**, 033003 (2012)
35. van den Broek, B., et al.: 2D Mater. **1**, 021004 (2014)
36. Quhe, R., et al.: Sci. Rep. **2**, 853 (2012)
37. Houssa, M., et al.: Appl. Phys. Lett. **98**, 223107 (2011)
38. Kane, C.L., Mele, E.J.: Phys. Rev. Lett. **95**, 226801 (2005)
39. Bernevig, B.A., et al.: Science **314**, 1757 (2006)
40. Liu, C.C., et al.: Phys. Rev. B **84**, 195430 (2011)
41. Zu, Y., et al.: Phys. Rev. Lett. **111**, 136804 (2013)
42. Zhu, F., et al.: Nat. Mater. **14**, 1020 (2015)
43. Konig, M., et al.: Science **318**, 766 (2012)
44. Houssa, M., et al.: J. Phys.: Condens. Matter **27**, 253002 (2015)
45. De Padova, P., et al.: J. Phys.: Condens. Matter **24**, 223001 (2012)
46. Chiappe, D., et al.: Adv. Mater. **24**, 5088 (2012)
47. Cinquanta, E., et al.: J. Phys. Chem. C **117**, 16719 (2013)
48. Jamgotchian, H., et al.: J. Phys.: Condens. Matter **24**, 172001 (2012)
49. Moras, P., et al.: J. Phys.: Condens. Matter **26**, 185001 (2014)
50. Vogt, P., et al.: Appl. Phys. Lett. **104**, 021602 (2014)
51. Grazianetti, C., et al.: J. Phys.: Condens. Matter **27**, 255005 (2015)
52. Molle, A., et al.: Adv. Funct. Mater. **23**, 4340 (2013)
53. Cinquanta, E., et al.: Phys. Rev. B **92**, 165427 (2015)
54. Scalise, E., et al.: 2D Mater. **1**, 011010 (2014)
55. Chiappe, D., et al.: Adv. Mater. **26**, 2096 (2014)
56. Molle, A., et al.: Adv. Mater. Interf. **3**, 1500619 (2016)

Decorated and Modified Graphenes as Electrodes in Na and Li-Ion Batteries

D. Pontiroli, G. Magnani, M. Gaboardi, M. Riccò, C. Milanese, J.C. Pramudita and N. Sharma

Abstract Nowadays, rechargeable Li-ion batteries represent the state of the art for the power supply in technological devices. However, the wide-scale implementation of this technology, for example in the automotive field or for large stationary applications, could raise issues, i.e. concerning the limited lithium mineral reserves. The investigation of alternatives to lithium is hence highly desirable, although it requires the identification of new materials suitable as components for new batteries, displaying similar or possibly even better performances with respect to the current systems. Here we show that electrodes based on graphene derivatives are able not only to support the insertion of Li^+, but also of Na^+ ions, with high capacity and stability upon cycling, leading to the development of novel Na-ion batteries.

1 Introduction

Electrochemical energy storage is one of the most promising strategies to get rid of fossil fuels; it may allow the reduction of intermittence of renewable energy power plants and the replacement of thermal engines with electric ones in vehicles [15]. The current leading technology in this field is represented by rechargeable Li-ion batteries (LIBs), which consist of two lithium insertion electrodes separated by a lithium ion conducting electrolyte, thus free from lithium metal, for safety reasons. The energy is stored thanks to the difference of free energy of lithium ions in the

D. Pontiroli · G. Magnani · M. Gaboardi · M. Riccò (✉)
Carbon Nanostructures Laboratory, Dipartimento di Scienze Matematiche, Fisiche e Informatiche, Università di Parma, Viale delle Scienze 7/a, 43124 Parma, Italy
e-mail: mauro.ricco@fis.unipr.it

C. Milanese
Pavia Hydrogen Lab, C.S.G.I. & Dipartimento di Chimica, Sezione di Chimica Fisica, Università di Pavia, Viale Taramelli 16, 27100 Pavia, Italy

J.C. Pramudita · N. Sharma
School of Chemistry, UNSW Australia, Sydney, NSW 2052, Australia

V. Morandi and L. Ottaviano (eds.), *GraphITA*, Carbon Nanostructures,
DOI 10.1007/978-3-319-58134-7_11

two electrodes and it is converted into electrical energy by letting flow lithium ions from one electrode to the other, through the electrolyte, thus feeding current in a device. This process can be reversed by an external current, able to restore the initial state of charge, and can be repeated many times before the battery runs out. LIBs offer high energy density and relatively long life cycle, but they still do not meet the demands of all current and emerging applications [23].

In fact, on the one hand, batteries are still not competitive with hydrocarbons in terms of volumetric and gravimetric energy/power density. Battery performances grew slowly in the last decades, as compared to other technologies, and it seems that LIBs may be reaching the limits of their abilities [17]. On the other hand, another serious limit in LIBs technology is represented by the relative abundance of lithium in the Earth's crust, which is only 20 ppm (see Fig. 1) and which would strongly affect the costs of LIBs if their utilisation would become extensive for large-scale applications [2]. For this reason, possible alternatives to lithium have been identified, which exploit more abundant elements, ranging from sodium, to divalent magnesium, to even trivalent aluminium [1, 13, 22]. In particular, sodium ion batteries (SIBs) are now increasingly being investigated thanks to the practically unlimited resources of sodium on Earth and to the similar chemistry shared by sodium and lithium. This would allow for parallel implementation of the materials/techniques developed for LIBs technology onto SIBs technology [16].

Furthermore, SIBs turn out to be more stable when discharged to lower potentials, showing better safety features as compared with LIBs [6]. Surprisingly, one of the main reasons of the lack of significant investments in SIBs so far probably has historical origin, since at the beginning of the big rush of LIBs research, after 1980, performances of Li systems were wrongly believed to be much higher as compared with Na systems [27]. Indeed, SIBs generally display lower energy densities than

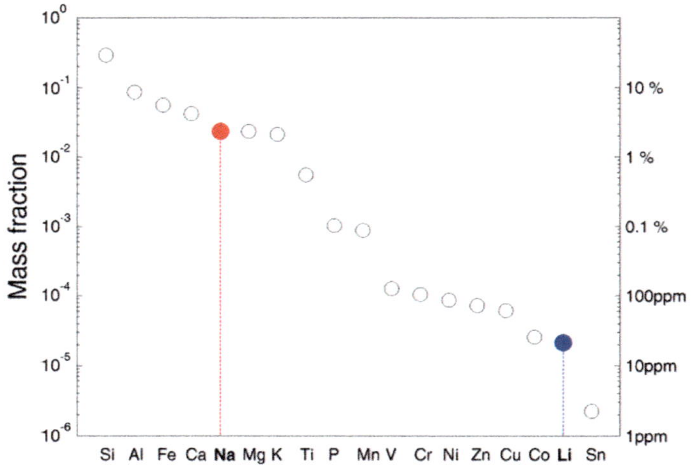

Fig. 1 Elemental abundance in the earth's crust [2]

LIBs, basically due to the decrease of Na reduction potential versus H_2/Pt and to the size increase of sodium relative to lithium, which limits ion diffusion. Nevertheless, these drawbacks would not be serious issues if SIB technology is applied, for example, in stationary energy storage systems, for which the mass is not a severe hinder, as compared, for example, with reliability. The current SIB technology identified some cathode materials as appropriate positive electrodes, e.g. $NaFePO_4$ and $Na_3V_2(PO_4)_3$ (corresponding to Li iron phosphate and Li vanadium phosphate in Li cells), capable to deliver theoretical capacities of 150 and 120 mAh/g, respectively [16, 26], but it still lacks a suitable anode material. This is because either graphite or silicon, which are the current anode materials used for LIBs, demonstrated to have poor performance when implemented into a sodium-ion system [22].

Several studies have investigated graphene in LIBs, which have shown significant improvement in battery performance. Graphene was used either as negative electrode [12], or was modified to form composites [4], or again it was used as a coating layer on anode and cathode materials [9, 11]. Use of graphene-based electrodes in SIBs has also started to emerge recently. Theoretical studies indicated a maximum capacity of 1450 mAh/g for graphene used as a negative electrode in a sodium-ion cell, even if no reversibility studies were performed [5].

In this work, we performed an electrochemical investigation of chemically produced graphene and several selected derivatives as electrodes in both Li- and Na-ion batteries, in particular hydrogenated graphene and nickel nanoparticles decorated graphene. It is worth noting that the synthesis of graphene-based materials used in this work allows obtaining large-scale (grams) amount of single and few-layer graphene and it is easily scalable to an industrial level, thus being remarkable for potential applications.

2 Sample Preparation and Characterisation

Graphene (TEGO) was produced by thermal exfoliation of graphite oxide. Graphite oxide (GO) was initially prepared following the Brodie method, discussed elsewhere [21], starting from highly pure graphite powder (SGL Carbon, RW-A grade). Specifically, TEGO was obtained by a sudden exposure of GO, sealed in a quartz vial, in a furnace at 1423 K, under dynamical vacuum. A sizeable expansion of the powder occurred as a result of the production of gases originated from the detachment of functional groups (see Fig. 2). This process also leads to the production of in-plane defects, in form of vacancies and internal edges, due to the removal of some carbon atoms from the graphene layers [21]. Hydrogenation of TEGO (H-TEGO) was performed by annealing TEGO at 1073 K under H_2 flux at ambient pressure for 1 h in a quartz vial. Decoration of TEGO with Ni nanoparticles (Ni-NPs) was obtained by the impregnation of TEGO with $Ni(acac)_2$ (Sigma Aldrich, 95% purity) in anhydrous tetrahydrofuran (THF, inhibitor free, Sigma Aldrich, 99.9% purity). This was followed by the complete removal of solvent and

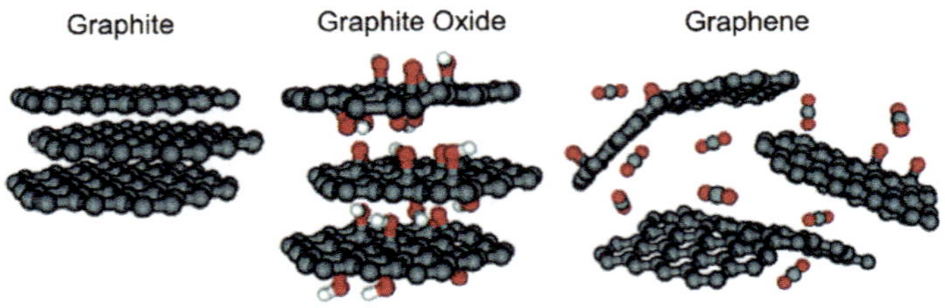

Fig. 2 Bulk production of graphene through the thermal exfoliation of graphite oxide. Reprinted with permission from [14]. Copyright 2007, American Chemical Society

by the thermal decomposition of $Ni(acac)_2$ heating up the sample at 573 K in dynamic vacuum (ramp rate 60 °C/h), thus leading to Ni-TEGO. It is worth noting that TEGO, H-TEGO and Ni-TEGO have never been exposed to air during their preparation and handling.

Previous characterisations (not shown here) indicated that TEGO consists of highly defective single and few-layer (typically three) graphene sheets, in which nanometer-sized perfect graphene islands are separated by more disordered carbon regions [8]. Defects originates principally by highly reactive carbon vacancies produced during the exfoliation process, and possibly from small residual GO regions. Such defects can subsequently be saturated by treating TEGO with hydrogen in H-TEGO; this treatment leads to a sizeable increase of the amount of H in the sample, which was successfully localised in correspondence of defects by NMR and inelastic neutron scattering investigation [19]. Typical diameter of monodispersed Ni-NPs in Ni-TEGO was found in the range 10–25 nm depending on Ni stoichiometry (in this study the Ni/C ratio is 1:80) and on the rate of the Ni $(acac)_2$ decomposition process. No clear traces of NiO thick shells on Ni-NPs were observed, either with transmission electron microscopy or by magnetometric investigation [8].

The electrodes were manufactured by mixing 70 wt% graphene-based active material with 30 wt% polyvinylidenefluoride binder (PVDF, MTI Corporation). A few mL of N-methylpyrrolidone (NMP, MTI Corporation) were added and the obtained slurry was magnetically stirred overnight. Then the slurry was deposited onto a copper foil by using a notch-bar and the electrode film was dried at 373 K in a vacuum oven overnight. The electrodes were pressed by using a flat-plate press (MTI Corporation) and dried under vacuum again at 373 K, they were subsequently transferred to an Ar glove box. Research coin cells (CR 2032) were constructed with graphene-based electrodes, with Li or Na metal counter-electrodes, separated by a polyethylene separator with 1 M $LiPF_6$ in dimethyl carbonate and diethyl carbonate (1:1 vol.%), or glass-fibre separator with 1 M $NaPF_6$ in dimethyl carbonate and diethyl carbonate (1:1 wt%), for LIBs and SIBs respectively.

3 Electrochemical Performances

The maximum first discharge capacity of LIBs with TEGO-based electrodes cycled at 100 mA/g reached the noticeable value of 3070 mAh/g, while the second discharge capacity suddenly decreased to 677 mAh/g. After 20 cycles the capacity decreased further to 300 mAh/g and stabilised up to the 50th cycle to a final value of 276 mAh/g (see Fig. 3a). The rate capability of TEGO, obtained cycling the cell at different applied current rates, resulted in a reasonable good capacity of 210 mAh/g at 1 mA/g current rate and 160 mAh/g at 5 mA/g, with a full recovery of the capacity at lower current rates after cycling at high rates (see Fig. 3b).

Electrodes of H-TEGO in LIBs showed a significant improvement of electrochemical performances, with respect to TEGO. In fact, a first capacity discharge of 6032 mAh/g was found, which decreased to 1351 mAh/g at the second discharge. On the 20th cycle the capacity decreased further to 600 mAh/g and stabilised at 488 mAh/g after 50 cycles (see Fig. 3c). A similar robustness as TEGO against high current rates was found for H-TEGO electrodes, as shown in Fig. 3d, with a less than 10% decrease in capacity for every increment of 1 mA/g of current.

On the other hand, Ni-TEGO displayed relatively poor performances when compared with TEGO and H-TEGO in LIBs. In fact, a breakdown of Ni-TEGO batteries was detected when cycled above 1.5 V, suggesting that adverse reactions occur at this potential (see Fig. 4a). This fact clearly limits the capacity performances of such cells, apparently in contradiction with recent results found in the literature.

In the case of SIBs, we found that remarkably the performance of TEGO, differently from graphite, does not decrease significantly. We found a first discharge capacity of 3040 mAh/g, which drops by 86% to 437 mAh/g at the second discharge, at 100 mA/g rate (see Fig. 3e). The capacity then stabilised to 248 mAh/g at the 50th cycle. High current rates of 1 A/g and 5 A/g resulted in capacity of 220 mAh/g and 175 mAh/g respectively, suggesting that TEGO is tolerant for high current rate cycling also in SIBs (see Fig. 3f).

Similarly to LIBs, H-TEGO electrodes in SIBs showed an improvement in performance compared with TEGO. An initial discharge capacity of 4072 mAh/g was measured, which decreased to 784 mAh/g at the second discharge. Capacity then stabilised to 491 mAh/g after 20 cycles, with coulombic efficiency of the order of 90%, as shown in Fig. 3g. At 1 A/g, capacity turn out to be 285 mAh/g, whereas cycling at 5 A/g it stabilised at 222 mAh/g. Also in case of H-TEGO good cyclability and recovery after high rates were detected (see Fig. 3h).

Finally, also Ni-TEGO showed a significant improvement in terms of performance as compared to TEGO, displaying a first discharge capacity of 6843 mAh/g, which decreased to 690 mAh/g after the second discharge. The capacity stabilised to 420 mAh/g after 20 cycles with 97% coulombic efficiency, as shown in Fig. 4b. The rate capabilities of Ni-TEGO showed a 21% of capacity drop at the rate of 1 A/g and a 46% capacity drop at 5 A/g relative to 100 mA/g, confirming a very good robustness of Ni-TEGO against high rates.

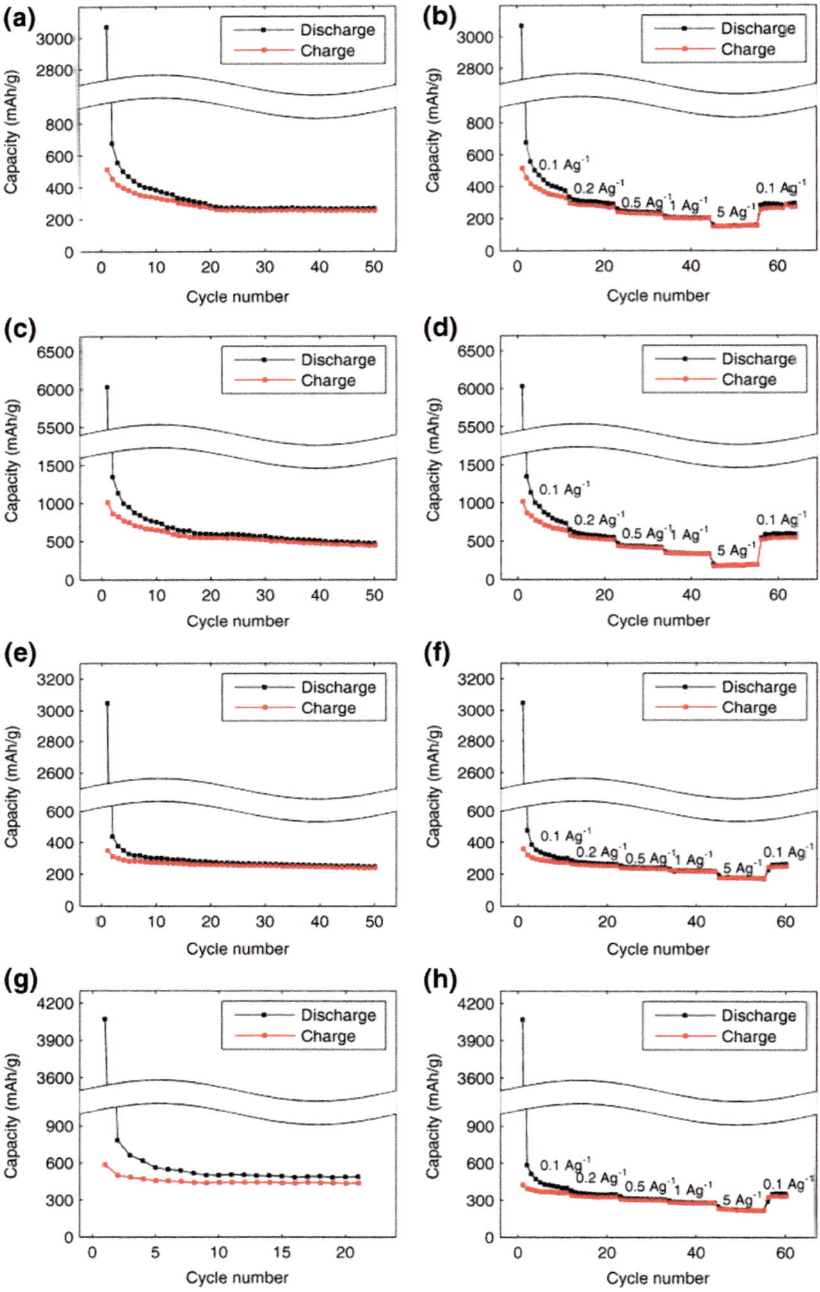

Fig. 3 Capacity versus cycle number for: **a** TEGO in LIBs, **c** H-TEGO in LIBs, **e** TEGO in SIBs and **g** H-TEGO in SIBs. The cells were cycled at 100 mA/g. Capacity versus cycle number for **b** TEGO in LIBs, **d** H-TEGO in LIBs, **f** TEGO in SIBs and **h** H-TEGO in SIBs cycled at variable current rates

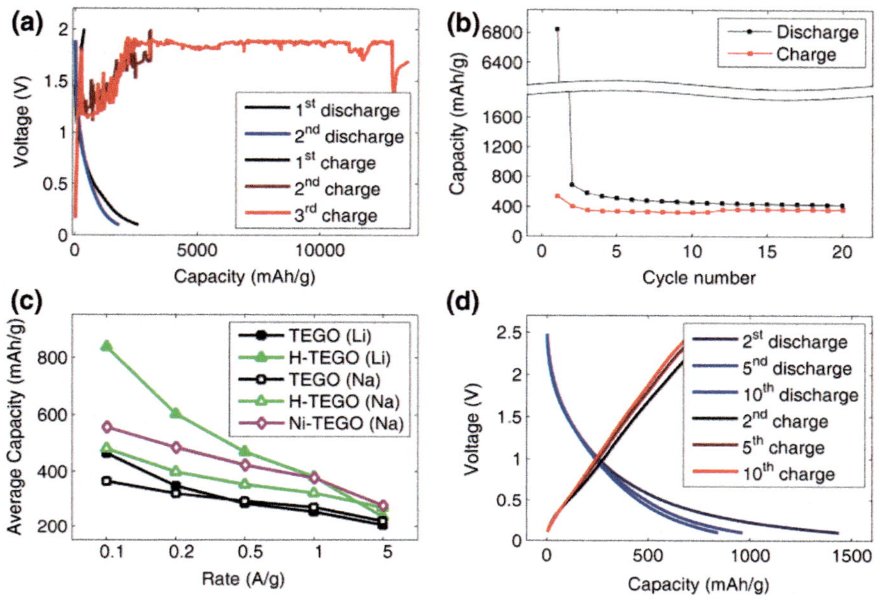

Fig. 4 **a** Potential versus capacity for Ni-TEGO in LIBs cycled above 1.5 V at 100 mA/g. **b** Capacity versus cycle number for Ni-TEGO in SIBs cycled at 100 mA/g. **c** Average capacity of TEGO, H-TEGO and Ni-TEGO in LIBs and SIBs at different current rates. **d** Potential versus capacity for Ni-TEGO in SIBs cycled at 100 mA/g

It is worth here noting that the performance of graphene-based electrodes at high rates in SIBs matched or even surpassed that in LIBs, as shown in Fig. 4c, suggesting that in these conditions Na insertion/extraction is more facile than Li insertion/extraction.

4 Discussion

The obtained results clearly indicated that, as soon as carbon layers of graphite are separated via chemical methods to obtain bulk graphene, the electrochemical performance in terms of capacity, rate capability and reliability of such material used as electrode noticeably increase, either in lithium, or, remarkably, even in sodium systems. Further modification of graphene revealed to allow the improvement of these properties, even if they do not work equally well for both battery systems. As far as LIBs are concerned, TEGO performance is clearly superior with respect to graphite [7], but also with respect to $Li_4Ti_5O_2$ and to Sn, especially in term of cyclability [10, 24]. On the other hand, in SIBs TEGO derivatives behave either better than hard carbon [18], or than TiO_2 nanoparticles [25].

In situ synchrotron diffraction measurements performed on TEGO in SIBs (not shown) excluded the possible reorganisation of graphene layers to form graphite during discharge/charge [20], which could explain the major drawback of these electrodes, namely the large irreversible capacity loss observed in the first cycle. On the other hand, the high specific surface of TEGO and TEGO derivatives (measured BET for TEGO resulted in about 500 m^2/g) probably plays an important role in the alkali intercalation mechanism. A capacitive behaviour seems to occur in charge/discharge cycles following the first cycle (see Fig. 4d), where a sloping potential versus capacity was observed for all the TEGO based electrodes. The high porosity of such materials is probably also the responsible of the large irreversible capacity loss observed in the first cycle, probably due to Li and Na trapping or SEI formation.

The significant improvement of electrochemical performances after the treatment with hydrogen in H-TEGO could be rationalised with the saturation of the high reactive in-plane defects produced during the exfoliation process, which are probably still active in TEGO. Inelastic neutron spectroscopy performed both on TEGO and H-TEGO (not shown) indicated that hydrogen is mainly localised at the external edges (borders) and in proximity of internal edges of large inner voids of graphene layers [3]. These extended carbon vacancies could possibly support the alkali diffusion also between inter-graphenic galleries, even if further investigation is required to better clarify the intercalation mechanism in these systems.

A sizeable enhancement of the capacity was observed in Ni-TEGO based SIBs, as compared with cells made with TEGO electrodes. On the other hand, the same decoration seems to be of detrimental in LIBs. This suggests that Ni-NPs should play an active role in the alkali intercalation process, even if the mechanism behind this behaviour is still unknown. Electron scanning microscopy (not shown) performed on Ni-TEGO electrodes in SIBs at different rate of charge indicated the presence of micrometre-size Na particles, which are not present in TEGO electrodes, suggesting that Ni-NPs could act as nucleation centres for Na aggregates [20], but further studies are required to better clarify the intercalation process.

5 Conclusions

We have shown that chemically produced graphene and some of its derivatives display very good electrochemical performance if used as negative electrodes in lithium ion batteries, in terms of either discharge capacity, rate capability, or recovery. Noticeably, the same materials in sodium ion batteries obtained similar performance. These last results are particularly attractive, since SIBs implementation and diffusion are still currently quite limited, due to the lack of identification of suitable negative electrodes so far, since common materials used in LIBs, such as graphite, do not intercalate Na appreciably. The relatively large irreversible capacity loss detected during the first discharge/charge cycle in such systems could be compensated by the pre-lithiation or by the pre-sodiation of the electrodes in the device.

Acknowledgements The authors would like to acknowledge the financial support from the Cariplo foundation (Project number 2013–0592, "Carbon based nanostructures for innovative hydrogen storage systems") and IRSES-EU Project MagNonMag nr. 295180.

References

1. Aurbach, D., Lu, Z., Schechter, A., Gofer, Y., Gizbar, H., Turgeman, R., Cohen, Y., Moshkovich, M., Levi, E.: Prototype systems for rechargeable magnesium batteries. Nature **407**(6805), 724–727 (2000)
2. Carmichael, R.S.: Practical handbook of physical properties of rocks and minerals. CRC Press, Boca Raton (FL) (1989)
3. Cavallari, C., Pontiroli, D., Jimenez-Ruiz, M., Johnson, M., Aramini, M., Gaboardi, M., Parker, S.F., Ricco, M., Rols, S.: Hydrogen motions in defective graphene: the role of surface defects. Phys. Chem. Chem. Phys. **18**(36), 24820–24824 (2016)
4. Choi, J., Jin, J., Jung, I.G., Kim, J.M., Kim, H.J., Son, S.U.: SnSe2 nanoplate-graphene composites as anode materials for lithium ion batteries. Chem. Commun. **47**(18), 5241–5243 (2011)
5. Datta, D., Li, J., Shenoy, V.B.: Defective graphene as a high-capacity anode material for Na- and Ca-Ion batteries. ACS Appl. Mater. Interfaces **6**(3), 1788–1795 (2014)
6. Datta, M.K., Epur, R., Saha, P., Kadakia, K., Park, S.K., Kumta, P.N.: Tin and graphite based nanocomposites: potential anode for sodium ion batteries. J. Power Sour. **225**, 316–322 (2013)
7. Fan, C.-L., Chen, H.: Preparation, structure, and electrochemical performance of anodes from artificial graphite scrap for lithium ion batteries. J. Mater. Sci. **46**(7), 2140–2147 (2010)
8. Gaboardi, M., Bliersbach, A., Bertoni, G., Aramini, M., Vlahopoulou, G., Pontiroli, D., Mauron, P., Magnani, G., Salviati, G., Zuttel, A., Ricco, M.: Decoration of graphene with nickel nanoparticles: study of the interaction with hydrogen. J. Mater. Chem. A **2**(4), 1039–1046 (2014)
9. Hassoun, J., Bonaccorso, F., Agostini, M., Angelucci, M., Betti, M.G., Cingolani, R., Gemmi, M., Mariani, C., Panero, S., Pellegrini, V., Scrosati, B.: An advanced lithium-ion battery based on a graphene anode and a lithium iron phosphate cathode. Nano Lett. **14**(8), 4901–4906 (2014)
10. Kim, Y.-J., Park, M.-S., Sohn, H.-J., Lee, H.: Electrochemical behaviors of SnO and Sn anodes for lithium rechargeable batteries. J. Alloy. Compd. **509**(12), 4367–4371 (2011)
11. Kucinskis, G., Bajars, G., Kleperis, J.: Graphene in lithium ion battery cathode materials: a review. J. Power Sour. **240**, 66–79 (2013)
12. Lian, P., Zhu, X., Liang, S., Li, Z., Yang, W., Wang, H.: Large reversible capacity of high quality graphene sheets as an anode material for lithium-ion batteries. Electrochim. Acta **55** (12), 3909–3914 (2010)
13. Lin, M.-C., Gong, M., Lu, B., Wu, Y., Wang, D.-Y., Guan, M., Angell, M., Chen, C., Yang, J., Hwang, B.-J., Dai, H.: An ultrafast rechargeable aluminium-ion battery. Nature **520**(7547), 324–328 (2015)
14. McAllister, M.J., Li, J.-L., Adamson, D.H., Schniepp, H.C., Abdala, A.A., Liu, J., Herrera-Alonso, M., Milius, D.L., Car, R., Prud'homme, R.K., Aksay, I.A.: Single sheet functionalized graphene by oxidation and thermal expansion of graphite. Chem. Mater. **19** (18), 4396–4404 (2007)
15. Owens, B.: Batteries. Nature **526**(7575), S89–S89 (2015)
16. Palomares, V., Serras, P., Villaluenga, I., Hueso, K.B., Carretero-Gonzalez, J., Rojo, T.: Na-ion batteries, recent advances and present challenges to become low cost energy storage systems. Energy Environ. Sci. **5**(3), 5884–5901 (2012)

17. Paradiso, J.A., Starner, T.: Energy scavenging for mobile and wireless electronics. IEEE Pervasive Comput. **4**(1), 18–27 (2005)
18. Ponrouch, A., Goñi, A.R., Palacín, M.R.: High capacity hard carbon anodes for sodium ion batteries in additive free electrolyte. Electrochem. Commun. **27**, 85–88 (2013)
19. Pontiroli, D., Aramini, M., Gaboardi, M., Mazzani, M., Sanna, S., Caracciolo, F., Carretta, P., Cavallari, C., Rols, S., Tatti, R., Aversa, L., Verucchi, R., Riccò, M.: Tracking the hydrogen motion in defective graphene. J. Phys. Chem. C **118**(13), 7110–7116 (2014)
20. Pramudita, J.C., Pontiroli, D., Magnani, G., Gaboardi, M., Riccò, M., Milanese, C., Brand, H. E.A., Sharma, N.: Graphene and selected derivatives as negative electrodes in sodium- and lithium-ion batteries. ChemElectroChem **2**(4), 600–610 (2015)
21. Riccò, M., Pontiroli, D., Mazzani, M., Choucair, M., Stride, J.A., Yazyev, O.V.: Muons probe strong hydrogen interactions with defective graphene. Nano Lett. **11**(11), 4919–4922 (2011)
22. Slater, M.D., Kim, D., Lee, E., Johnson, C.S.: Sodium-Ion Batteries. Adv. Func. Mater. **23**(8), 947–958 (2013)
23. Tarascon, J.M., Armand, M.: Issues and challenges facing rechargeable lithium batteries. Nature **414**(6861), 359–367 (2001)
24. Wang, J., Zhao, H., Wen, Y., Xie, J., Xia, Q., Zhang, T., Zeng, Z., Du, X.: High performance $Li_4Ti_5O_{12}$ material as anode for lithium-ion batteries. Electrochim. Acta **113**, 679–685 (2013)
25. Wu, L., Buchholz, D., Bresser, D., Gomes Chagas, L., Passerini, S.: Anatase TiO_2 nanoparticles for high power sodium-ion anodes. J. Power Sour. **251**, 379–385 (2014)
26. Xiong, H., Slater, M.D., Balasubramanian, M., Johnson, C.S., Rajh, T.: Amorphous TiO_2 nanotube anode for rechargeable sodium ion batteries. J. Phys. Chem. Lett. **2**(20), 2560–2565 (2011)
27. Yabuuchi, N., Kubota, K., Dahbi, M., Komaba, S.: Research development on sodium-ion batteries. Chem. Rev. **114**(23), 11636–11682 (2014)

Chemically Exfoliated Layered Materials for Practical Gas Sensing Applications

F. Perrozzi, C. Cantalini and L. Ottaviano

Abstract In the last decade two-dimensional materials have attracted great interest for a wide range of applications, from optoelectronics to nanobiotechnology. Between the possible applications, the use of two-dimensional materials for the production of gas sensors has proved to be very promising. In this work we report on the fabrication and the characterization of gas sensors obtained from chemically exfoliated graphene oxide, black phosphorous, MoS_2 and WS_2. All the sensors were prepared by simply drop casting the flakes dispersed in a solution on a Pt interdigitated substrate. The sensors have shown excellent electrical responses to both oxidizing and reducing gases, demonstrating that the use of chemically exfoliated two-dimensional materials is a practical and cheap way to produce high-quality gas sensing devices.

1 Introduction

Since the discovery of graphene (G) [1], two-dimensional (2D) materials have attracted great interest for gas sensing application thanks to their high surface to mass ratio and their wide range of chemical and physical properties [2]. Beside graphene, the most studied and promising 2D materials for such applications are graphene oxide (GO) [3–6], black phosphorus (BP) [7, 8] and the transition metal dichalcogenides (TMDs) [9] such as MoS_2 [10, 11] and WS_2 [12–14]. Depending on the 2D materials production technique different kind of sensors have been fabricated [2].

F. Perrozzi (✉) · L. Ottaviano
Department of Physical and Chemical Sciences, University of L'Aquila,
Via Vetoio 10, 67100 L'Aquila, Italy
e-mail: francesco.perrozzi@gmail.com

C. Cantalini
Department of Industrial and Information Engineering and Economics,
University of L'Aquila, Via G. Gronchi 18, 67100 L'Aquila, Italy

V. Morandi and L. Ottaviano (eds.), *GraphITA*, Carbon Nanostructures,
DOI 10.1007/978-3-319-58134-7_12

The first sensors were prepared by mechanically exfoliating the starting material and creating monolayer flake devices with lithographic techniques. Single-layer devices have demonstrated detection limits of individual gas molecules for graphene [15] and few ppb to NO_2 for BP [7, 16]. Excellent sensing performances have been demonstrated also for MoS_2 [10, 17–19] and WS_2 [12]. If these kind of sensors represent the best way to study the sensing properties of the 2D materials, they are not easy to make and, above all, their production is not scalable. Another way to produce gas sensors is by chemical vapor deposition (CVD). CVD gas sensors have shown excellent sensitivity [20, 21] and their production is scalable. The main drawbacks of these sensors are the cost and the technical production difficulties. Chemical exfoliation techniques for 2D materials production, on the other hand, allows to build cheap but high-quality chemoresistive sensors simply by wet depositing the flakes dispersed in a solution on an interdigitated substrate [4, 5, 8, 11, 14, 22].

In this work we report on the fabrication, the characterization and the sensing properties of gas sensors obtained by drop casting chemically exfoliated GO, BP, MoS_2 and WS_2 on an interdigitated substrate. We have carefully analyzed the thermal and chemical stability of our sensors that, in order to overcome the limitations related to the slow gas desorption cinematics, usually work at operating temperatures (OT) higher that room temperature (RT). The sensors have been tested and annealed at a temperature of 150–200 °C, RT, 150–250 °C, and 150–450 °C respectively for GO, BP, MoS_2, and WS_2. The analysis of the chemical and structural changes induced by the exfoliation procedures and by the annealing is fundamental to understand the gas sensing mechanism of the devices. The sensors have shown good sensing performances with a remarkable low detection limit to NO_2 of 20 ppb for GO [4], BP [8] and MoS_2 [11] devices and a low detection limit to H_2 and NH_3 of 1 ppm for the WS_2 device [14].

2 Materials and Methods

2.1 Materials Production

GO is prepared via a modified Hummers method [23] starting from graphite flakes of 500 μm maximum size. Monolayers with ~30 μm average size are obtained and dispersed in water for eventual deposition.

To obtain exfoliated BP flakes [8, 24] 20 mg of high purity (99.99%) BP crystal pieces (Smart-elements, GmbH) have been dispersed in 4 ml of anhydrous N-methylpyrrolidone (NMP, Sigma-Aldrich). The dispersion has been sonicated for 5 h by an ultrasonic equipment (UP100H) controlled at 25 °C, set at 100 W and 30 kHz. To avoid BP degradation due to solvated H_2O, the sonication process has been performed in a glow box, under N_2 controlled atmosphere (T = 29.2 °C, RH = (4.5 ± 2)%).

MoS_2 exfoliated flakes were obtained by sonicating a 7.5 mg/ml solution of MoS_2 powder (Sigma–Aldrich, CAS 1317-33-5) into N-methyl-pyrrolidone (NMP) (Sigma–Aldrich, CAS 872-50-4) for 1 h [11, 25].

WS_2 thin films were prepared by drop casting commercial WS_2 nano-flakes dispersed in ethanol/water solution purchased from Graphene Laboratory inc (www.graphene-supemarket.com). The purity and the concentration of the solution were 99% and 26 mg/l respectively. The lateral size of WS_2 flakes (1–4 mono-layers) was 50–150 nm [14].

2.2 Sensors Fabrications

All the sensors were prepared depositing the solution on 30 μm spaced interdigitated Pt electrodes patterned on Si_3N_4 substrate. The GO sensor was prepared by drop casting 80 μL of 0.1 mg/mL GO solution and air drying at 50 °C. BP sensor has been fabricated in the same glow box under N_2 atmosphere by drop casting 20 μL of the as prepared solution and then stabilized under N_2 at 75 °C for 3 h. The MoS_2 sensor was prepared by drop casting 50 μL of the NMP solution and heated in air to stabilize (pressure 1 atm, humidity 50%) at 50 °C for 10 min. The WS_2 sensors were prepared by drop casting the solution and then the sensor was oven-annealed in static air for 1 h within the temperature range 150–350 °C (heating and cooling rate of 5 °C/min).

2.3 Sensors Characterization

The chemical composition of the sensors was measured by XPS technique. XPS spectra have been acquired with a PHI 1257 spectrometer equipped with a monochromatic Al Kα source ($h\nu = 1486.6$ eV) with a pass energy of 11.75 eV (93.9 eV survey), corresponding to an overall experimental resolution of 0.25 eV. The acquired XPS spectra have been fitted with Voigt line shapes and Shirley backgrounds.

The dispersion of the flakes on the patterned substrate and their lateral size were checked with a Scanning Electron Microscope (SEM, model ZEISS LEO-1430).

The electrical resistance of the films in H_2, NH_3, and NO_2 gases was measured by an automated system. Thin films deposited on Si_3N_4 provided with Pt finger type electrodes were placed inside a Teflon chamber (500 cm^3), provided with Teflon tubing and exposed to gaseous mixtures of H_2/NH_3 and NO_2 in dry air carrier at 500 sc cm/min flow rate. The average gas residence time inside the chamber is set at 1 min. Different gas concentrations were obtained by mixing certified H_2, NH_3, and NO_2 mixtures (250 ppm H_2, 250 ppm NH_3 and 5 ppm NO_2 in dry air cylinder) with dry air carrier, by means of an MKS147 multi gas mass controller. NO_2 gas concentrations in the downstream were measured by means of Ansyco AS32 M chemiluminescence analyzer (Environment S.A). The sensors electrical resistance was measured by means of the volt-amperometric technique (AGILENT 34970A) at different OTs in the temperature range 25–350 °C by heating the Si_3N_4 substrate through backside-circuit *dc* current injections (20–80 mA). In the analysis of the gas responses, the sensitivity is defined as the slope of the output calibration curve that is sensor response intensity versus gas concentration. The response intensity is defined as the ratio between the electrical resistance to a target gas (R_g) and the electrical resistance in air (R_a) for a given gas concentration if $R_g > R_a$ or vice versa if $R_a > R_g$. The adsorption and desorption times are defined as the time required to reach 90% of the whole gas response at equilibrium and of the base line resistance (i.e., the resistance in dry air) respectively.

3 Results and Discussion

3.1 Gas Sensors Structure

In Fig. 1a it is reported the basics of the device fabrication. Thirty micrometer-spaced inter-digitated Pt electrodes are patterned on Si_3N_4 substrate. Figure 1b shows a picture of the patterned texture (front-size) with heating elements and temperature sensors on the backside. In order to work a connection between two

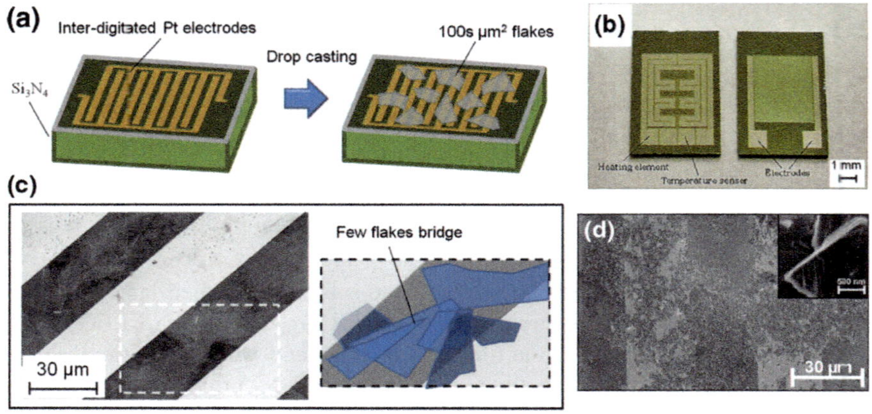

Fig. 1 **a** Sketch of the device fabrication. Electric contacts between pre-patterned interdigitated Pt electrodes are realized upon drop casting deposition of large GO flakes. **b** Patterned substrate (front-size) with heating elements and temperature sensors on the backside. **c** SEM image of few GO flakes bridging two adjacent Pt electrodes. Reproduced with permission from [4]. Copyright 2013, American Chemical Society. **d** SEM image of the MoS_2 flakes deposited onto Si_3N_4 substrate with pre-patterned Pt interdigitated electrodes (bright stripes). Reproduced with permission from [11]. Copyright 2014, Elsevier B.V

adjacent electrodes must be established. Depending on the average size of the flakes of the 2D material, the bridge between the electrodes can be made with few or several flakes.

In Fig. 1c we report a SEM image of the GO sensor where the electric contact is obtained with few or one GO flakes. This is possible tanks to the large size of GO flakes that can measure hundreds μm in diameter. The schematic view helps the identification of the flakes in the dashed line marked area [4].

In the case of exfoliated BP [8], MoS_2 [11], and WS_2 [14] the lateral size of the flakes is much lower than the inter-electrodes distance (30 μm), in a way that adjacent electrodes cannot be connected using just one or few flakes, as for GO. In Fig. 1d is reported the SEM image of MoS_2 sensor [11]. The figure reveals that the drop casting deposition does not allow a uniform coverage of the substrate. However, as demonstrated also for the BP and WS_2 sensors [8, 14], many bridging sites have been formed from the union of several flakes, thus creating percolation pathways throughout the sensors surface. This 'patchwork architecture' affects the baseline resistance of the devices, but, as discussed later on, not theirs gas sensing performances.

3.2 Thermal and Chemical Stability

Thermal and chemical stability of the sensors have been investigated by XPS technique. All the sensors, with the exception of the BP one, have been annealed to improve the time response and the electrical base line recovery. In the case of GO the annealing is necessary to reduce it and to make it conductive, while for the BP the sensor was not heated and used only in dry air in order to avoid oxidation and degradation [26, 27].

In Fig. 2 we report the XPS C1s core-level spectra of as-deposited GO (a) and 200 °C UHV annealed GO (b). The GO analysis has been extended to GO annealed at 200 °C in ultrahigh vacuum to check the evolution of the chemical bonds in the OT range (25–200 °C) in the absence of contaminants. Both C1s spectra are fitted by the sum of five components assigned to C atoms belonging to aromatic rings and hydrogenated carbon ($\underline{C}$ = C/C–C, $\underline{C}$–H, 284.6–284.9 eV), hydroxyl groups ($\underline{C}$–OH, 285.9 eV), epoxy groups ($\underline{C}$–O–C, 286.9 eV), carbonyl groups ($\underline{C}$ = O, 288.0 eV), and carboxyl groups ($\underline{C}$ = O(OH), 289.3 eV) [28, 29]. As a consequence of thermal annealing, the groups containing oxygen tend to disappear, unlike those containing aromatic carbon, indicating a loss of oxygen in favor of sp^2 carbon domains formation. The analysis of the GO annealed at 150 °C in air, i.e., the OT of the GO sensor, is reported in [4]. There it is shown that the XPS spectra of the 150 °C annealed GO is similar to the 200 °C UHV one, demonstrating that the effects of mild annealing are not affected by the presence of air contaminants.

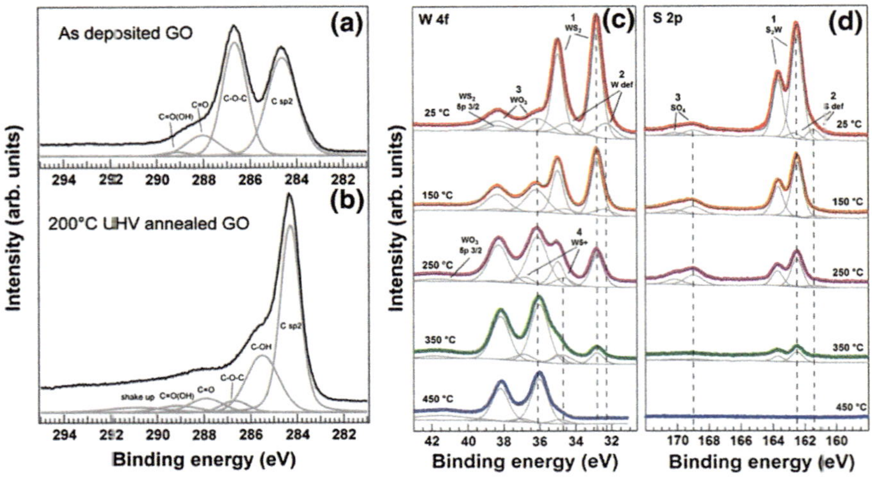

Fig. 2 XPS C1s core-level spectra of as-deposited GO **a** and 200 °C UHV annealed GO **b** Reproduced with permission from [4]. Copyright 2013, American Chemical Society. W 4f **c** and the S 2p **d** core-level photoemission spectra acquired on WS_2 thin films before (25 °C) and after annealing in air at 150, 250, 350 and 450 °C. All the W 4f (S 2p) components are composed by the $4f_{7/2}$ ($2p_{3/2}$) and $4f_{5/2}$ ($2p_{1/2}$) shifted peaks. Reproduced with permission from [14]. Copyright 2016, Elsevier B.V

The BP XPS analysis reveals that the exfoliation procedure in controlled N_2 atmosphere and that the operating conditions at RT and in dry air of the sensor prevent the formation of any PO_x species [8].

For the MoS_2 and the WS_2 the XPS analysis shows that the main effect of the annealing in air is the progressive oxidation of the flakes [11, 14]. In Fig. 2c, d are reported the evolution of the W 4f and the S 2p core-level photoemission spectra of the WS_2 sensor as a function of the annealing temperature respectively. The W 4f core-level spectra are characterized by four chemically shifted components, each of which composed by the W $4f_{7/2}$ and the W $4f_{5/2}$ doublet, assigned to W belonging to hexagonal WS_2 (1 WS_2) and to defective WS_2 (2 W def), and to W fully coordinated (3 WO_3) and partially coordinated (4 W 5 +) in WO_3 [14]. The S 2p core-level spectra are characterized by three components, each composed by the S $2p_{3/2}$ and the S $2p_{1/2}$, assigned to S belonging to hexagonal WS_2 (1 S_2W) and to defective WS_2 (2 S def), and to S belonging to SO_4^{2-} anions (3 SO_4) [14]. Similar evolution of the XPS spectra has been reported for the MoS_2 [11], with some notable differences related to the solvent used for the chemical exfoliation. Indeed, it has been demonstrated [11] that the presence of intercalated NMP can dope the MoS_2 with nitrogen atoms, and thus affecting its electronic properties.

The XPS analysis demonstrates that, when annealed, the starting material undergoes a change in terms of reduction/oxidation, and in terms of structural defects production. The presence of oxidation and defects in the devices is supposed to play a major role in their gas sensing mechanism [4, 8, 11, 14].

3.3 Sensing Properties

In Fig. 3 are reported the GO, WS_2, and MoS_2 sensors response to NO_2 or H_2/NH_3 acquire at their optimal operating temperature, i.e. the temperature representing the best compromise in terms of conductivity, sensitivity limits and stability of the sensors.

Figure 3a reports the GO response at 150 °C for different NO_2 concentrations. The device resistance is monitored as a function of the time. The NO_2 concentration is increased from 20 ppb to 10 ppm with steps of 1 h exposure to NO_2 (adsorption) and 1 h exposure to dry air (desorption). The bottom-left inset shows the detection limit of 20 ppb. The corresponding sensitivity curve (0.5 slope) is reported in the

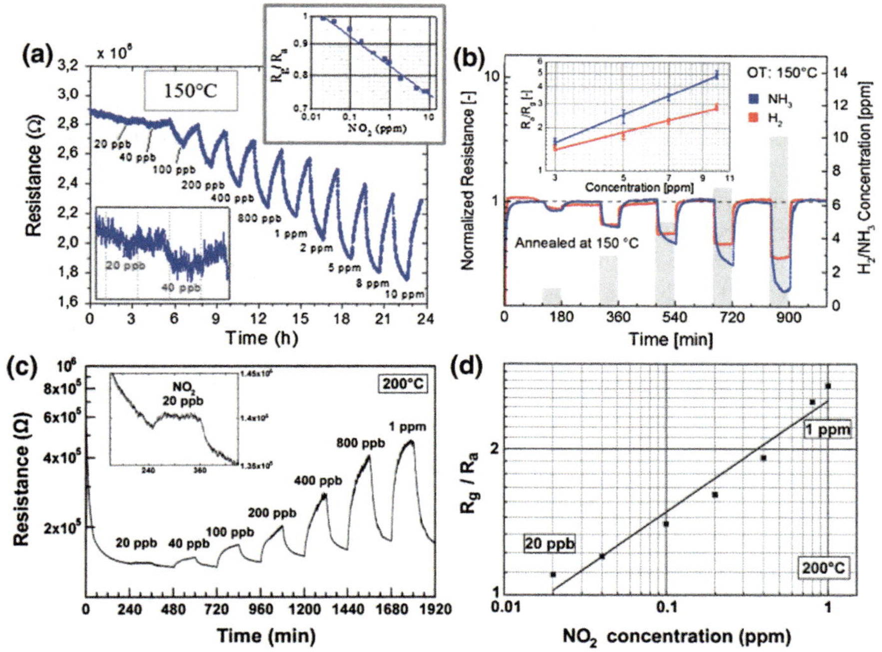

Fig. 3 **a** GO response to increasing NO_2 concentration at 150 °C OT. Device resistance versus time is reported in the main panel. *Bottom-left inset*: zoom in the detection limit range. *Top-right inset*: sensitivity curve. Reproduced with permission from [4]. Copyright 2013, American Chemical Society. **b** Electrical responses of the WS_2 sensor at 150 °C OT exposed to identical concentration of H_2 and NH_3 (1–10 ppm) in dry air carrier. The *inset* shows the calibrating lines to NH_3 and H_2 respectively. The bars in the inset represent the standard deviation calculated over a set of 5 measurements performed for each gas concentration. Reproduced with permission from [14]. Copyright 2016, Elsevier B.V. **c** Dynamic response in dry air of the MoS_2 device to NO_2 in the range 20 ppb–1 ppm at 200 °C OT. *Inset*: zoom-in of the electrical response to 20 ppb NO_2. **d** Calibration curve of the MoS_2 device at 200 °C OT in dry air. A linear interpolation curve is reported. The sensor sensitivity is the slope of the calibration curve. Reproduced with permission from [11]. Copyright 2014, Elsevier B.V

top-right inset. The very low detection limit can be ascribed to the high number of active oxygen sites which, at the same time, have a detrimental effect on the response time (few minutes at 150 °C OT versus the few seconds of highly reduced GO-based devices [30]). Additional measurements of GO gas sensing properties with respect to other gases have been performed. In particular GO-based gas sensor has shown a detection limit of 100 ppm to H_2, and has shown to be sensitive to NH_3, toluene, ethanol, and acetone in conditions of gas saturation [5]. Finally, the GO sensor is characterized by a *p*-type semiconductor behavior once exposed to both oxidative (NO_2) and reducing (H_2, NH_3) target gases [4, 5].

Figure 3b shows the electrical responses of the WS_2 sensor at 150 °C OT, exposing the film to same concentrations of reducing H_2 and NH_3 gases in dry air carrier. A low detection limit of 1 ppm H_2 and 1 ppm NH_3 are clearly displayed. The inset reports the calibrating lines to H_2 and NH_3 which show that the sensor has a higher sensitivity to NH_3 as compared to H_2, but with bigger adsorption and desorption response times. Furthermore, the WS_2 sensor response in the presence of NO_2 has been analyzed and it has shown a low detection limit of 40 ppb. The WS_2 sensor is characterized by a *n*-type semiconductor behavior [14]. The comparison between the 150, 250, and 350 °C annealed WS_2 sensors shows that the presence of WO_3 does not influences the gas sensing responses at 150 °C, while it became important at higher temperatures [14].

Figure 3c reports the dynamic response of the MoS_2 device operating at 200 °C to NO_2 in the range 20–1000 ppb. The inset clearly shows that the sensor has a low detection limit of 20 ppb NO_2. Given the observed quality of the signal to noise ratio in Fig. 3c, it is easy to believe that even lower detection limits could be measured. Figure 3d reports the calibrating plot of the device to NO_2, calculated from the experimental results of Fig. 3c. The slope of the fitting line represents the sensitivity of the device, and it is equal to 0.23 ± 0.01. Furthermore, the MoS_2 sensor has shown a good and sharp response to 250 ppm H_2. The comparison between the 150 °C annealed sensor and the 250 °C annealed one demonstrates that the solvent used in the exfoliation procedure, i.e., the NMP, can modify the electronic properties of the MoS_2 by doping it with N atoms. The doping N atoms can be removed by annealing the MoS_2 at temperatures higher than 150–200 °C. The consequence of doping is that the 150 °C MoS_2 sensor is characterized by a *p*-type response, while the 250 °C annealed sensor has shown a *n*-type semiconductor behavior, similarly to the WS_2 one [11]. For both WS_2 and MoS_2 the excellent gas sensing performances are explained by the presence of defects such as S vacancies produced by the exfoliation procedure and the thermal annealing [11, 14].

The BP sensor, once electrically stabilized at 75 °C in N_2 atmosphere, has been used at RT and in dry air because of its well-known chemical instability [26, 27]. The gas responses of the device to increasing concentrations of NH_3, H_2 and NO_2 shows a low detection limit of 20 ppm, 5 ppm and 20 ppb respectively. The main drawbacks of the BP sensor are the slow time response and the incomplete recovery of the base line after gas exposure, especially after the exposure to NH_3. Both these problems are a consequence of the low OT of the device. The BP sensor is characterized by a *p*-type response [8].

Table 1 Gas sensing low detection limit measured for GO, BP, MoS_2 and WS_2 devices

	NO_2 (ppb)	H_2 (ppm)	NH_3
GO (150 °C)	20	100	–
BP (RT)	<20	<5	<10 ppm
MoS_2 (200 °C)	<20	<250	–
WS_2 (150 °C)	100	1	1 ppm

In Table 1 are summarized the low detection limits to NO_2, H_2, and NH_3 of the chemically exfoliated 2D materials gas sensors used.

4 Conclusion

In conclusion we have demonstrated that the simple deposition of chemically exfoliated 2D materials on interdigitated substrates is a good way to produce practical, cheap, and high performances gas sensing devices. Remarkable low detection limits of 20 ppb to NO_2 for the GO, BP and the MoS_2 sensors and of 1 ppm to H_2 and NH_3 for the WS_2 sensor have been reported. The sensing mechanisms and the devices performances have been analyzed in light of the chemical and structural modifications induced by the exfoliation procedure and the thermal annealing.

References

1. Novoselov, K.S., Jiang, D., Schedin, F., Booth, T.J., Khotkevich, V.V., Morozov, S.V., Geim, A.K.: Two-dimensional atomic crystals. Proc. Natl. Acad. Sci. U.S.A. **102**(30), 10451–10453 (2005). doi:10.1073/pnas.0502848102
2. Yang, W., Gan, L., Li, H., Zhai, T.: Two-dimensional layered nanomaterials for gas-sensing applications. Inorg. Chem. Front. **3**(4), 433–451 (2016). doi:10.1039/c5qi00251f
3. Yuan, W., Liu, A., Huang, L., Li, C., Shi, G.: High-performance NO_2 sensors based on chemically modified graphene. Adv. Mater. **25**(5), 766–771 (2013). doi:10.1002/adma.201203172
4. Prezioso, S., Perrozzi, F., Giancaterini, L., Cantalini, C., Treossi, E., Palermo, V., Nardone, M., Santucci, S., Ottaviano, L.: Graphene oxide as a practical solution to high sensitivity gas sensing. J. Phys. Chem. C. **117**(20), 10683–10690 (2013). doi:10.1021/Jp3085759
5. Donarelli, M., Prezioso, S., Perrozzi, F., Giancaterini, L., Cantalini, C., Treossi, E., Palermo, V., Santucci, S., Ottaviano, L.: Graphene oxide for gas detection under standard humidity conditions. 2D Mater. **2**(3) (2015). doi:10.1088/2053-1583/2/3/035018
6. Perrozzi, F., Prezioso, S., Ottaviano, L.: Graphene oxide: from fundamentals to applications. J. Phys. Condens. Matter. **27**(1) (2015). doi:10.1088/0953-8984/27/1/013002
7. Abbas, A.N., Liu, B.L., Chen, L., Ma, Y.Q., Cong, S., Aroonyadet, N., Kopf, M., Nilges, T., Zhou, C.W.: Black phosphorus gas sensors. Acs Nano. **9**(5), 5618–5624 (2015). doi:10.1021/acsnano.5b01961

8. Donarelli, M., Ottaviano, L., Giancaterini, L., Fioravanti, G., Perrozzi, F., Cantalini, C.: Exfoliated black phosphorus gas sensing properties at room temperature. 2D Mater. **3**(2) (2016). doi:10.1088/2053-1583/3/2/025002
9. Lv, R., Robinson, J.A., Schaak, R.E., Sun, D., Sun, Y.F., Mallouk, T.E., Terrones, M.: Transition metal dichalcogenides and beyond: synthesis, properties, and applications of single- and few-layer nanosheets. Acc. Chem. Res. **48**(1), 56–64 (2015). doi:10.1021/Ar5002846
10. Li, H., Yin, Z., He, Q., Li, H., Huang, X., Lu, G., Fam, D.W.H., Tok, A.I.Y., Zhang, Q., Zhang, H.: Fabrication of single- and multilayer MoS 2 film-based field-effect transistors for sensing NO at room temperature. Small **8**(1), 63–67 (2012). doi:10.1002/smll.201101016
11. Donarelli, M., Prezioso, S., Perrozzi, F., Bisti, F., Nardone, M., Giancaterini, L., Cantalini, C., Ottaviano, L.: Response to NO2 and other gases of resistive chemically exfoliated MoS2-based gas sensors. Sens. Actuators B-Chem. **207**, 602–613 (2015). doi:10.1016/j.snb.2014.10.099
12. Huo, N., Yang, S., Wei, Z., Li, S.S., Xia, J.B., Li, J.: Photoresponsive and gas sensing field-effect transistors based on multilayer WS 2 nanoflakes. Sci. Rep. **4** (2014). doi:10.1038/srep05209
13. Kuru, C., Choi, D., Kargar, A., Liu, C.H., Yavuz, S., Choi, C., Jin, S., Bandaru, P.R.: High-performance flexible hydrogen sensor made of WS_2 nanosheet-Pd nanoparticle composite film. Nanotechnology **27**(19) (2016). doi:10.1088/0957-4484/27/19/195501
14. Perrozzi, F., Emamjomeh, S.M., Paolucci, V., Taglieri, G., Ottaviano, L., Cantalini, C.: Thermal stability of WS_2 flakes and gas sensing properties of WS_2/WO_3 composite to H_2, NH_3 and NO_2. Sens. Actuators B: Chem. **243**, 812–822 (2017). doi:10.1016/j.snb.2016.12.069
15. Schedin, F., Geim, A.K., Morozov, S.V., Hill, E.W., Blake, P., Katsnelson, M.I., Novoselov, K.S.: Detection of individual gas molecules adsorbed on graphene. Nat. Mater. **6**(9), 652–655 (2007). doi:10.1038/Nmat1967
16. Cui, S., Pu, H., Wells, S.A., Wen, Z., Mao, S., Chang, J., Hersam, M.C., Chen, J.: Ultrahigh sensitivity and layer-dependent sensing performance of phosphorene-based gas sensors. Nat. Commun. **6** (2015). doi:10.1038/ncomms9632
17. He, Q., Zeng, Z., Yin, Z., Li, H., Wu, S., Huang, X., Zhang, H.: Fabrication of flexible MoS2 thin-film transistor arrays for practical gas-sensing applications. Small **8**(19), 2994–2999 (2012). doi:10.1002/smll.201201224
18. Late, D.J., Huang, Y.K., Liu, B., Acharya, J., Shirodkar, S.N., Luo, J.J., Yan, A.M., Charles, D., Waghmare, U.V., Dravid, V.P., Rao, C.N.R.: Sensing behavior of atomically thin-layered MoS2 transistors. ACS Nano **7**(6), 4879–4891 (2013). doi:10.1021/Nn400026u
19. Perkins, F.K., Friedman, A.L., Cobas, E., Campbell, P.M., Jernigan, G.G., Jonker, B.T.: Chemical vapor sensing with monolayer MoS2. Nano Lett. **13**(2), 668–673 (2013). doi:10.1021/nl3043079
20. Kim, Y.H., Kim, S.J., Kim, Y.J., Shim, Y.S., Kim, S.Y., Hong, B.H., Jang, H.W.: Self-activated transparent all-graphene gas sensor with endurance to humidity and mechanical bending. ACS Nano **9**(10), 10453–10460 (2015). doi:10.1021/acsnano.5b04680
21. Kumar, S., Kaushik, S., Pratap, R., Raghavan, S.: Graphene on paper: a simple, low-cost chemical sensing platform. ACS Appl. Mater. Interfaces. **7**(4), 2189–2194 (2015). doi:10.1021/am5084122
22. Zhang, S.L., Choi, H.H., Yue, H.Y., Yang, W.C.: Controlled exfoliation of molybdenum disulfide for developing thin film humidity sensor. Curr. Appl. Phys. **14**(3), 264–268 (2014). doi:10.1016/j.cap.2013.11.031
23. Treossi, E., Melucci, M., Liscio, A., Gazzano, M., Samorì, P., Palermo, V.: High-contrast visualization of graphene oxide on dye-sensitized glass, quartz, and silicon by fluorescence quenching. J. Am. Chem. Soc. **131**(43), 15576–15577 (2009). doi:10.1021/ja9055382
24. Nicolosi, V., Chhowalla, M., Kanatzidis, M.G., Strano, M.S., Coleman, J.N.: Liquid exfoliation of layered materials. Science **340**(6139),1420– + (2013). doi:10.1126/Science.1226419

25. Coleman, J.N., Lotya, M., O'Neill, A., Bergin, S.D., King, P.J., Khan, U., Young, K., Gaucher, A., De, S., Smith, R.J., Shvets, I.V., Arora, S.K., Stanton, G., Kim, H.Y., Lee, K., Kim, G.T., Duesberg, G.S., Hallam, T., Boland, J.J., Wang, J.J., Donegan, J.F., Grunlan, J.C., Moriarty, G., Shmeliov, A., Nicholls, R.J., Perkins, J.M., Grieveson, E.M., Theuwissen, K., McComb, D.W., Nellist, P.D., Nicolosi, V.: Two-dimensional nanosheets produced by liquid exfoliation of layered materials. Science **331**(6017), 568–571 (2011). doi:10.1126/science.1194975
26. Castellanos-Gomez, A., Vicarelli, L., Prada, E., Island, J.O., Narasimha-Acharya, K.L., Blanter, S.I., Groenendijk, D.J., Buscema, M., Steele, G.A., Alvarez, J.V., Zandbergen, H.W., Palacios, J.J., van der Zant, H.S.J.: Isolation and characterization of few-layer black phosphorus. 2D Mater. **1**(2) (2014). doi:10.1088/2053-1583/1/2/025001
27. Wood, J.D., Wells, S.A., Jariwala, D., Chen, K.S., Cho, E., Sangwan, V.K., Liu, X., Lauhon, L.J., Marks, T.J., Hersam, M.C.: Effective passivation of exfoliated black phosphorus transistors against ambient degradation. Nano Lett. **14**(12), 6964–6970 (2014). doi:10.1021/nl5032293
28. Stankovich, S., Dikin, D.A., Piner, R.D., Kohlhaas, K.A., Kleinhammes, A., Jia, Y., Wu, Y., Nguyen, S.T., Ruoff, R.S.: Synthesis of graphene-based nanosheets via chemical reduction of exfoliated graphite oxide. Carbon **45**(7), 1558–1565 (2007). doi:10.1016/j.carbon.2007.02.034
29. Perrozzi, F., Prezioso, S., Donarelli, M., Bisti, F., De Marco, P., Santucci, S., Nardone, M., Treossi, E., Palermo, V., Ottaviano, L.: Use of optical contrast to estimate the degree of reduction of graphene oxide. J. Phys. Chem. C. **117**(1), 620–625 (2013). doi:10.1021/jp3069738
30. Robinson, J.T., Perkins, F.K., Snow, E.S., Wei, Z., Sheehan, P.E.: Reduced graphene oxide molecular sensors. Nano Lett. **8**(10), 3137–3140 (2008). doi:10.1021/nl8013007

Solutions of Reduced Carbon Allotropes and Their Utilization for Functional Material Generation

F. Hof and A. Pénicaud

Abstract Despite their reactive nature, the properties of solutions of reduced carbon allotropes have been studied in detail and remarkable research regarding their possible utilization have been conducted in recent years. Solutions of reduced carbon allotropes are an excellent basis for the generation of functional architectures. Solutions of reduced carbon nanotubides have been used to enrich metallic SWCNTS in solution, to fabricate transparent conductive films and to generate macroscopic fibers of aligned CNTs. GICs have been used to produce potassium batteries and to synthesize carbon nanodots with photoluminescence behavior. We foresee a bright future for these solutions because predominantly single components are present, e.g., single-layer graphene and their properties are now well understood.

1 Introduction

Research on nanometer-sized carbon allotropes has attracted worldwide interest in related fields such as chemistry, physics, and material science [1]. This can be related to the fascinating properties of these materials, which are based on the structural blueprint of the carbon allotropes. A broad variety of different characteristics can be found, i.e., from metallic conductivity up to isolating behavior. The promising properties of these materials prompted countless research initiative, to pave the way for these unique materials in next generation applications. Especially, the 2D material graphene has attracted special interest in recent years, due to its high conductance, transparency, tensile strength, and chemical stability [1–4].

Two distinct different routes are currently used to obtain single layered graphene, the bottom up and top down approach. In both fields of research, tremendous efforts have been invested to gain significant progress to obtain high-quality graphene in

F. Hof · A. Pénicaud (✉)
CNRS, Centre de Recherche Paul Pascal (CRPP), UPR 8641, 33600 Pessac, France
e-mail: penicaud@crpp-bordeaux.cnrs.fr

F. Hof · A. Pénicaud
Université de Bordeaux, CRPP, UPR 8641, 33600 Pessac, France

V. Morandi and L. Ottaviano (eds.), *GraphITA*, Carbon Nanostructures,
DOI 10.1007/978-3-319-58134-7_13

recent years. Especially, liquid formulations of graphene but also those of novel, nano sized carbon allotropes are scientifically and technologically highly sought after. The top down approach in case of graphite is appealing because the starting material is abundant and can be exfoliated to single layered graphene in the course of the process. The most prominent methods to stabilize graphene or related carbon nano materials in liquid formulations are either surfactant-based methods, or oxidation reactions of the carbon framework. These topics have been reviewed intensively and are beyond the scope of this article [5, 6]. On the other hand, reductive dissolution allows an exhaustive exfoliation of carbon nanomaterials. In this article, the latest achievements in the field of reductive intercalation and solubilization of carbon allotropes will be presented. Especially, studies regarding the application of graphene intercalation compounds (GICs) will be discussed in detail but also recent developments in the field of carbon nanotubides will be presented. Each of the methods to generate liquid formulations of graphene have their strengths and their weaknesses, which are related to crucial limits like stability, content of truly single-layered material, reproducibility, scalability, and ease of preparation.

2 Discussion

The very first description of the intercalation of graphite by potassium was published in 1926 by Fredenhagen [7, 8] and further intercalation compounds have been described a few years later [9]. For almost a century now, these compounds have attracted interest in the research community because of their interesting structure, defined by stages, and their properties, e.g., superconductivity [10]. Nowadays, the intercalation of a variety of different, novel carbon allotropes such as fullerenes [11], carbon nanotubes (CNTs) [12], carbon nanohorns [13], and carbon nano-onions [14] have been accomplished and their properties studied. In general, the intercalation can be achieved by the direct reaction of the intercalant and the carbon allotrope under reduced pressure and/or elevated temperature, under inert gas atmosphere. An interesting calculation was made on the diffusion of alkali metals in GIC compounds. Special attention has been paid on the diffusion rates of metal atoms in between the carbon planes [15]. Diffusion constants of the alkali metals at 25 °C have been calculated and it was found that their diffusion constant increases as Li < Na < K. At 25 °C, values in the range of 10^{-10} m^2s^{-1} have been found for diffusion of potassium atoms between the carbon planes. Furthermore, promoters such as amines and naphthalene can be used to fine tune final concentrations or reaction parameters [16, 17]. Since the era of graphene, a renewed focus has been drawn on potassium graphite intercalation compounds as GICs have been recognized as readily exfoliable graphitic material. Consequently, these have been exploited to generate solutions of graphenide in aprotic, polar solvents. The generation of stable graphenide solutions can be rationalized by the structure of the GIC. Here, the large π–π interactions in graphite crystals are effectively replaced by electrostatic interactions. An almost complete charge transfer from the potassium to the graphite has been demonstrated for potassium-based GICs [18]. Furthermore, the interlayer distance is enlarged from 3.35 Å in graphite to ca 5.35 Å in KC_8 (stage I intercalation compound). This is common within all carbon-based intercalation compounds, as the respective π–π interactions are replaced by electrostatic interactions. The fact that there is an actual spontaneous dissolution of the GIC in certain organic solvents is connected to the Gibbs free energy where the energy of mixing and the entropy of mixing define if a solution in a thermodynamical sense can be generated. The stability of the final solution is influenced by the chemical nature of the carbon allotrope, the amount of charges, the nature of counter ions, and their respective distribution. Thus, a complex picture can be anticipated where the rigidity of the lattices as well as the respective shapes of the different carbon allotropes may result in either enhanced stability or decreased one in specific solvents. In consequence, for every intercalated carbon allotrope, the type of solvent, the amount of charge and the concentrations in a specific solvent have to be optimized to maximize the stability of the generated solution. A more comprehensive discussion can be found in our previous review and book chapter [19, 20].

The particular advantage of intercalated carbon allotropes is the possibility to achieve exfoliation of the material down to its single component (graphene in the case of graphite). For CNTs [21] as well as for graphene [22] it has been demonstrated that upon dissolution of the respective intercalation compound, without any sonication treatment, predominantly single-layered material can be obtained [23, 24]. This finding is in a sharp contrast to surfactant-based methods, where first, sonication treatment or shear mixing is inevitable to disperse the carbon material,[1] accompanied by shortening/cutting of the respective carbon allotrope and second, a distribution of layered materials is obtained, with a maximum of few layered materials [26, 27]. This behavior and the observed distribution of layers in these systems can be explained by the strong π–π interactions in the starting carbon material that are replaced by solvent/surfactant/surface interaction in the case of water based dispersions. In the case of GIC solutions the π–π interactions of the carbon material have been replaced by electrostatic interactions. This conceptual difference explains the observations that single layer material can be easily obtained by dissolving GICs in organic solvents [28, 29], whereas the concentration of single-layer material in aqueous, surfactant-based environment is significantly lower.

The charge on these systems is on the one hand the reason why the intercalated carbons can be stabilized in solution, on the other hand it demands the preparation, handling and storage of these solution under exclusion of oxidizing agents such as oxygen and moisture. The charged carbon materials can be understood as reduction agent themself. Therefore, carbon intercalation compounds, especially potassium based ones, have been utilized to effectively functionalize the carbon framework in

[1]Sodium deoxycholate aqueous dispersions may constitute a notable exception: CNT individualization has been demonstrated with no sonication treatment albeit at the expense of long stirring times (several weeks) [25].

recent publications [13, 14, 30–34]. The interested reader is referred to the well written review by C. Jiang et al. that covers the reductive functionalization of CNTs in more detail [35] (Fig. 1).

The reactivity explains the difficulties retrieving information of these materials in situ, because special focus and care have to be applied to obtain reliable spectroscopic information from either the intercalated and/or the solvated species [36, 37]. Difficulties in the interpretation of the results can be encountered when partially or fully reoxidized systems are studied, as functionalization of the carbon framework has to be considered. This is directly related to one of the main challenges in the field of reduced carbon materials. It is the question of removing charges from the carbon framework without the introduction of functional groups [30]. Still today, an effective, mild, and easy procedure is highly desired to oxidize the respective carbon allotrope without any covalent functionalization to be able to fully utilize the extraordinary properties of the nanosized carbon materials [38, 39].

In the next paragraph the development of applications based on GICs will be highlighted. One specific, simultaneously interesting application is the development of batteries from GIC compounds. Although, lithium ion batteries are widespread nowadays, potassium ion batteries are rare. The first effective preparation of potassium GIC batteries has been reported recently [40]. In contrast to the common procedure to produce lithium ion batteries, in case of the potassium battery it is crucial to first prepare the potassium GIC and used this material in a second step as electrode materials in a formulation of water free electrolytes .

The presented procedure ensures that the carbon lattice interspacing is enlarged and the carbon is efficiently intercalated. The reported results are promising and the performance of the batteries may be increased, once a better and more stable electrolyte system is found. The synthesis of graphene-based quantum dots and carbon metal nanohybrids have also been reported. In the former case, GIC of small graphitic compounds have been generated. This material was exfoliated, transferred to water and their photoluminescence properties have been described [41]. Photoluminescence has been observed also in the solution of nanocarbon platelets generated by the dissolution of potassium intercalated carbon fibers [42]. Finally, GICs have been applied as a reduction agent to produce copper and nickel nanocomposites of graphene [43, 44] (Fig. 2).

In the last couple of years, a renewed focus was drawn on carbon nanotubides and consequently several studies have been published from different groups. This focus can be explained by improved spectroscopic techniques, availability of high-quality starting material and the development of separation techniques, based on soap dispersions, to obtain single chirality CNTs. One interesting work was published by Jiang et al. They have demonstrated that the concentration of the carbon nanotubide can be increased, when crown ether (18-crown-6) is added to the carbon nanotubide solution. Crown ether complexes the potassium cation and increases the stability of the solution. The maximum concentration of the carbon nanotubide in solution was so high, that liquid crystalline behavior was observed [45]. This is an interesting observation because it overcomes one of the main limitations of the reductive solubilization method which is the concentration of the

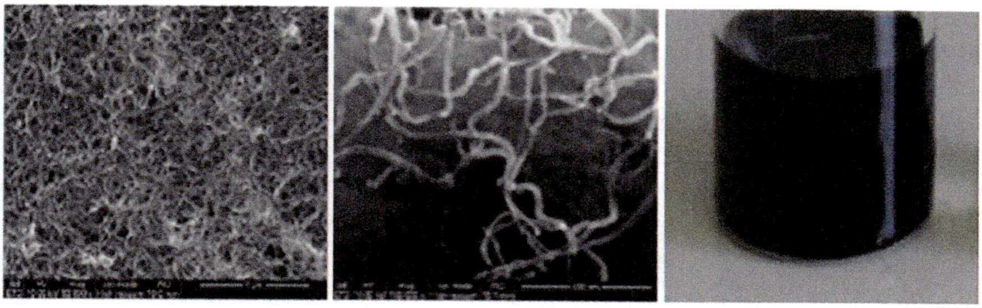

Fig. 1 *Left* and *middle* SEM images of multiwall carbon nanotube films prepared by drop casting of solutions of reduced multi wall carbon nanotubes. The density of the films can be controlled by the concentration of the MWCNTs in solution. *Right* Image of a concentrated DMSO solution of multi wall carbon nanotubes. Adapted from Ref. [19]

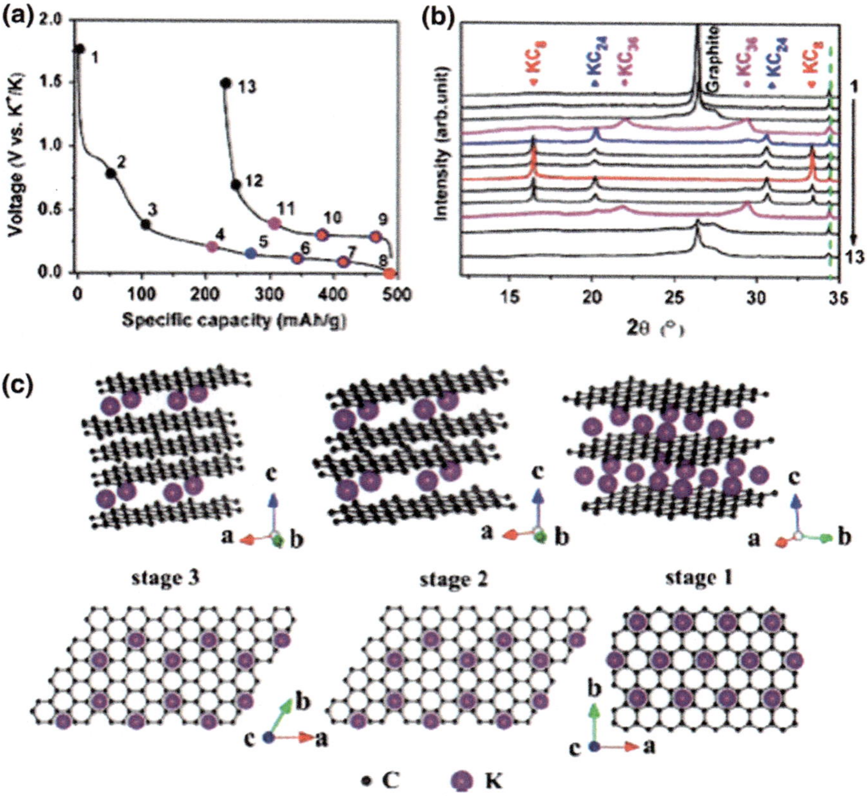

Fig. 2 **A** First cycle galvanostatic potassiation/depotassiation tests. **B** XRD data of the electrode materials. **C** Schematic presentation of the graphite intercalation compounds stage 3–1. Stage 1 compounds have the stoichiometric formula of KC_8. Adapted from Ref. [40]

respective solutions. Typically, the concentration of the solutions of reduced carbon allotropes is below 0.5 mg/mL. The electrochemical potential of carbon nanotubide solutions was studied by Hodge et al. [46]. In this study, metal salts have been added to the reduced carbon nanotubes and the measured metal content in these samples was determined by ICP-AES. By comparing measured values, a semi-quantitative measure of the density of states of the dissolved carbon nanotubide has been shown. In a recent study, it was demonstrated that by fine tuning the stoichiometry of the potassium to carbon ratio in carbon nanotubes using ratios less than 1/200 K/C a selective charging of metallic SWCNTs could be achieved. This result is a promising first step, to develop a method to efficiently separate metallic and semiconducting SWCNTs by the aid of reductive dissolution. The reason for the selective reduction was explained by the more favorable stabilization of charges in metallic carbon nanotubes than in semiconducting ones at low charging ratios. The selective charging was used to enrich metallic carbon nanotubes while dissolving in absolute THF [47] (Fig. 3).

A direct likewise interesting application of solutions of intercalated carbons is the fabrication of transparent conductive films (TCFs). Especially, carbon nanotubes are a promising alternative to replace metal oxide in conductive thin films as these films exhibit high conductivity and transparency. In addition, they are flexible as well as resistant to harsh chemical environment where ITO is not stable [48]. The solutions of carbon nanotubides have been exploited and one order of magnitude higher conductivity at 80% transmission have been demonstrated in comparison to aqueous dispersions of SWCNTs stabilized by the aid of sodium cholate [48]. Further improvements of the deposition technique, introducing orientation in the resulting SWCNT network by a shear force method, DC conductivities up to 7000 S cm^{-1} could be reached [49]. Carbon nanotube films generated from carbon nanotubide solutions have been also utilized to study electrochemiluminescence behavior [50]. As proof of concept, functionalized polystyrene microbeads have

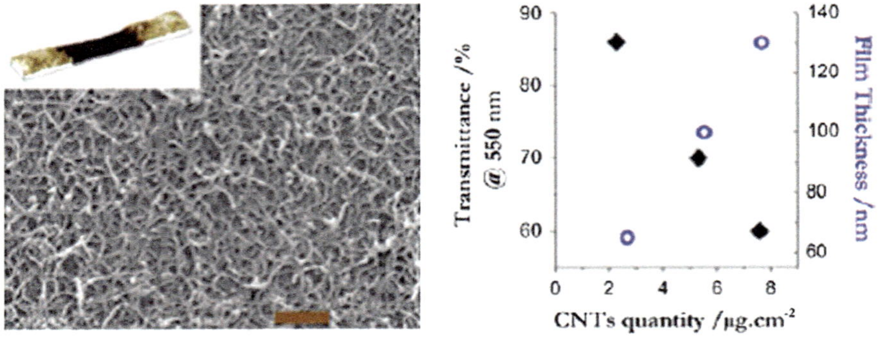

Fig. 3 *Left side* SEM image of a multiwall carbon nanotube films. *Inset* AFM image to determine the respective thickness of the film. *Right side* Plot of the transmittance at 550 nm and the thickness of the films in dependence of the CNT quantity. Adapted from Ref. [48]

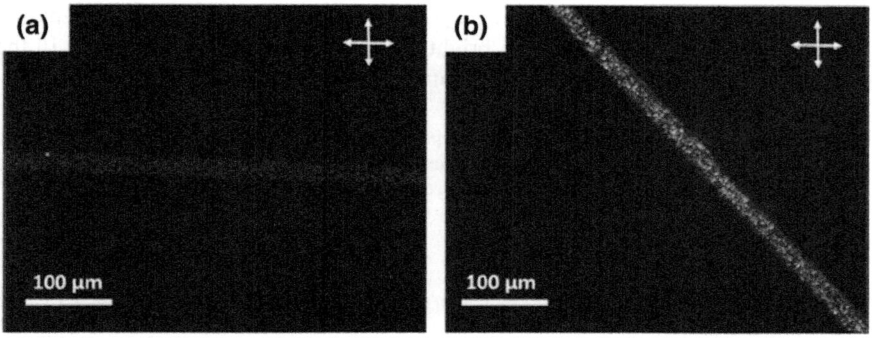

Fig. 4 Cross-polarized optical images of fibers made of DMSO solutions containing carbon nanotubides and crown ether. **a** The fiber axis is parallel to the polarizer. **b** The fiber axis is rotated at 45° to the polarizer. Images are adapted from Ref. [51]

been synthesized and used in electrochemical luminescence studies, highly interesting for bio-imaging.

Not only can TCFs but also macroscopic fibers can be generated from carbon nanotubide solutions. By utilizing alignment techniques and high carbon concentration (up to 60 mg/mL) in the 18-crown-6 polyelectrolyte, fibers with high tensile strength and conductive behavior could be produced (Fig. 4).

As outlined in the manuscript, this technique opens the possibility to direct cross-link SWCNTs and therefore also higher strength fibers are expected to be generated [51]. Recently, cyrogels have been fabricated utilizing DMSO-based solution of carbon nanotubes, which may lead to development of interesting architectures [24]. Finally, overcoming the limitation of inert atmosphere handling, it has been shown that graphenide THF solutions can be transferred into degassed water, to yield air stable suspensions of single-layered graphene in water after evaporation of THF [52–54].

3 Conclusions

The reductive dissolution of carbon allotropes has proven to be one of the most reliable methods to generate thermodynamically stable solutions, not only from graphene but also from various different nano sized carbon allotropes such as carbon nanotubes. The recently published results shine light on the mechanistic understanding of these systems and describe developments of applications that rely on the unique properties of these solutions. Recent outstanding trademarks can be summarized in the successful enrichment of metallic carbon nanotubes utilizing the respective charge, the fabrication of conductive thin films that exhibit superior conductivity and the producing of potassium ion batteries involving GIC compounds. Despite the challenging synthesis of the intercalation compounds of carbon-based materials, we see a bright future for these compounds for further groundbreaking research but also for commercial proposes and products.

Acknowledgements Funding from the European Union's Seventh Framework Program for research, technological development and demonstration under grant agreement No. 603488 (Plascarb project) is acknowledged. This work has been done within the framework of the GDR-I 3217 "graphene and nanotubes".

References

1. Georgakilas, V., Perman, J.A., Tucek, J., Zboril, R.: Chem. Rev. **115**, 4744 (2015)
2. Gogotsi, Y.: MRS Bull. **40**, 1110 (2015)
3. Ferrari, A.C., Bonaccorso, F., Fal'ko, V., Novoselov, K.S., Roche, S., et al.: Nanoscale **7**, 4598 (2015)
4. Garcia-Hernandez, M., Coleman, J.: 2D Mater. **3**, 010401 (2016)
5. Eigler, S., Hirsch, A.: Angew. Chemie - Int. Ed. **53**, 7720 (2014)
6. Nicolosi, V., Chhowalla, M., Kanatzidis, M.G., Strano, M.S., Coleman, J.N.: Science **340**, 1226419 (2013)
7. Fredenhagen, K.: J. Inorg. Gen. Chem. **158**, 249 (1926)
8. Fredenhagen, K., Suck, H.: Zeitschrift Für Anorg. Und Allg. Chemie **178**, 353 (1929)
9. Thiele, H.: Zeitschrift Für Anorg. Und Allg. Chemie **206**, 407 (1932)
10. Csányi, G., Littlewood, P.B., Nevidomskyy, A.H., Pickard, C.J., Simons, B.D.: Nat. Phys. **1**, 42 (2005)
11. Haddon, R.C., Hebard, A.F., Rosseinsky, M.J., Murphy, D.W., Duclos, S.J., et al.: Nature **350**, 320 (1991)
12. Rao, A.M., Eklund, P.C., Bandow, S., Thess, A., Smalley, R.E.: Nature **388**, 257 (1997)
13. Voiry, D.A., Pagona, G., Del Canto, E., Ortolani, L., Morandi, V., Noe, L., Monthioux, M., Tagmatarchis, N., Penicaud, A.: Chem. Commun. **51**, 5017 (2015)
14. Molina-Ontoria, A., Chaur, M.N., Plonska-Brzezinska, M.E., Echegoyen, L.: Chem. Commun. **49**, 2406 (2013)
15. Wang, Z., Ratvik, A.P., Grande, T., Selbach, S.M.: RSC Adv. **5**, 15985 (2015)
16. Liang, F., Sadana, A.K., Peera, A., Chattopadhyay, J., Gu, Z., Hauge, R.H., Billups, W.E.: Nano Lett. **4**, 1257 (2004)
17. Voiry, D., Vallés, C., Roubeau, O., Pénicaud, A.: Carbon N. Y. **49**, 170 (2011)
18. Grüneis, A., Attaccalite, C., Rubio, A., Vyalikh, D.V., Molodtsov, S.L., Fink, J., Follath, R., Eberhardt, W., Büchner, B., Pichler, T.: Phys. Rev. B - Condens. Matter Mater. Phys. **80**, 1 (2009)
19. Pénicaud, A.: Top. Curr. Chem. 1–35 (2014)
20. Pénicaud, A., Drummond, C.: Acc. Chem. Res. **46**, 129 (2013)
21. Pénicaud, A., Poulin, P., Derré, A., Anglaret, E., Petit, P.: J. Am. Chem. Soc. **127**, 8 (2005)
22. Vallés, C., Drummond, C., Saadaoui, H., Furtado, C.A., He, M., Roubeau, O., Ortolani, L., Monthioux, M., Pénicaud, A.: J. Am. Chem. Soc. **130**, 15802 (2008)
23. Milner, E.M., Skipper, N.T., Howard, C.A., Shaffer, M.S.P., Buckley, D.J., Rahnejat, K.A., Cullen, P.L., Heenan, R.K., Lindner, P., Schweins, R.: J. Am. Chem. Soc. **134**, 8302 (2012)
24. Pénicaud, A., Dragin, F., Pécastaings, G., He, M., Anglaret, E.: Carbon N. Y. **67**, 360 (2014)
25. Wenseleers, W., Vlasov, I.L., Goovaerts, E., Obraztsova, E.D., Lobach, A.S., Bouwen, A.: Adv. Funct. Mater. **14**, 1105 (2004)
26. Coleman, J.N.: Adv. Funct. Mater. **19**, 3680 (2009)

27. Lucas, A., Zakri, C., Maugey, M., Pasquali, M., van der Schoot, P., Poulin, P.: J. Phys. Chem. C **113**, 20599 (2009)
28. Catheline, A., Vallés, C., Drummond, C., Ortolani, L., Morandi, V., Marcaccio, M., Iurlo, M., Paolucci, F., Pénicaud, A.: Chem. Commun. (Camb.) **47**, 5470 (2011)
29. Catheline, A., Ortolani, L., Morandi, V., Melle-Franco, M., Drummond, C., Zakri, C., Pénicaud, A.: Soft Matter **8**, 7882 (2012)
30. Hof, F., Bosch, S., Eigler, S., Hauke, F., Hirsch, A.: J. Am. Chem. Soc. **135**, 18385 (2013)
31. Hof, F., Schäfer, R.A., Weiss, C., Hauke, F., Hirsch, A.: Chem. A Eur. J. **20**, 16644 (2014)
32. Clancy, A.J., Melbourne, J., Shaffer, M.S.P.: J. Mater. Chem. A **3**, 16708 (2015)
33. Maeda, Y., Takehana, Y., Yamada, M., Suzuki, M., Murakami, T.: Chem. Commun. (Camb.) **51**, 13462 (2015)
34. Vecera, P., Edelthalhammer, K., Hauke, F., Hirsch, A.: Phys. Status Solidi Basic Res. **251**, 2536 (2014)
35. Jiang, C., Saha, A., Martí, A.A.: Nanoscale **7**, 15037 (2015)
36. Tristant, D., Wang, Y., Gerber, I., Monthioux, M., Pénicaud, A., Puech, P.: J. Appl. Phys. **118**, 044304 (2015)
37. Wang, Y., Puech, P., Gerber, I., Pénicaud, A.: J. Raman Spectrosc. **45**, 219 (2014)
38. Hodge, S. et al.: Chemical routes to discharging graphenides, Nanoscale, **9**(9), 3150–3158, (2017)
39. Vecera, P. et al.: Solvent-driven electron trapping and mass transport in reduced graphites to access perfect graphene, Nat. comm. **7**, 12411, (2016)
40. Jian, Z., Luo, W., Ji, X.: J. Am. Chem. Soc. **137**, 11566 (2015)
41. Song, S.H., Jang, M.-H., Chung, J., Jin, S.H., Kim, B.H., Hur, S.-H., Yoo, S., Cho, Y.-H., Jeon, S.: Adv. Opt. Mater. **2**, 1016 (2014)
42. Huang, K., Delport, G., Orcin-Chaix, L., Drummond, C., Lauret, J.-S., Penicaud, A.: Nanoscale **8**, 8810 (2016)
43. Neiva, E.G.C., Souza, V.H.R., Huang, K., Pénicaud, A., Zarbin, A.J.G.: J. Colloid Interface Sci. **453**, 28 (2015)
44. Bayazit, M.K., Hodge, S.A., Clancy, A.J., Menzel, R., Chen, S., Shaffer, M.S.P.: Chem. Commun. (Camb). **52**, 1934 (2015)
45. Jiang, C., Saha, A., Xiang, C., Young, C.C., Tour, J.M., Pasquali, M., Martí, A.A.: ACS Nano **7**, 4503 (2013)
46. Hodge, S.A., Tay, H.H., Anthony, D.B., Menzel, R., Buckley, D.J., Cullen, P.L., Skipper, N. T., Howard, C.A., Shaffer, M.S.P.: Faraday Discuss. **172**, 311 (2014)
47. Gebhardt, J., Bosch, S., Hof, F., Hauke, F., Hirsch, A., Görling, A.: J. Mater. Chem. C (2016)
48. Catheline, A., Paolucci, F., Valenti, G., Poulin, P., Pénicaud, A.: J. Mater. Res. **30**, 2009 (2015)
49. Tune, D.D., Blanch, A.J., Shearer, C.J., Moore, K.E., Pfohl, M., Shapter, J.G., Flavel, B.S.: ACS Appl. Mater. Interfaces **7**, 25857 (2015)
50. Valenti, G., Zangheri, M., Sansaloni, S.E., Mirasoli, M., Penicaud, A., Roda, A., Paolucci, F.: Chem. - A Eur. J. **21**, 12640 (2015)
51. Jiang, C., Saha, A., Young, C.C., Hashim, D.P., Ramirez, C.E., Ajayan, P.M., Pasquali, M., Martí, A.A.: ACS Nano **8**, 9107 (2014)
52. Bepete, G., Anglaret, E., Ortolani, L., Morandi, V., Huang, K., Pénicaud, A., Drummond, C.: Nat. Chem. **9**, 347 (2017)
53. Bepete, G., Hof, F., Huang, K., Kampioti, K., Anglaret, E., Drummond, C., Pénicaud, A.: Phys. Status Solidi - Rapid Res. Lett. **10**, 895 (2016)
54. Bepete, G., Pénicaud, A., Drummond, C., Anglaret, E.: J. Phys. Chem. C **120**, 28204 (2016)

Synthesis of High-Density Graphene Foams Using Nanoparticle Templates

M. Christian, L. Venturi, L. Ortolani, F. Liscio, R. Rizzoli, V. Palermo and V. Morandi

Abstract Graphene foams grown by CVD on commercially available Ni foam templates were first reported in 2011. Since then they have been investigated widely due to their ability to transfer many of the unique properties of graphene to the macroscopic scale, with high surface area, high electrical conductivity and good structural integrity. However, the pore-size range is typically 200–400 μm, so much of the volume is unoccupied by the functional graphene material. We report a new synthesis procedure that produces graphene foams with pore sizes in the range of 1–10 μm, by using a sacrificial template of metal nanoparticles sintered together to form a network. These materials could have wide-ranging applications in fields such as high-density energy storage, membranes and sensing.

1 Introduction

Although graphene has innumerable useful properties, manipulating the 2D sheets into macroscopic structures is critical for many applications. Typically, 3D graphene structures such as papers [1] and hydrogels [2] are synthesised from chemically exfoliated graphene, but these tend to have relatively poor electrical conductivity because of their many defects and non-continuous nature. Instead, high-quality graphene may be grown on templates of virtually any shape by means of Chemical Vapour Deposition (CVD), and in this way free-standing macrostructures called graphene foams (GF) were first produced from nickel (Ni) foam templates in 2011

M. Christian (✉) · L. Venturi · L. Ortolani · F. Liscio · R. Rizzoli · V. Morandi
CNR-IMM Bologna, Via Gobetti 101, 40129 Bologna, Italy
e-mail: christian@bo.imm.cnr.it

M. Christian · V. Palermo
CNR-ISOF Bologna, Via Gobetti 101, 40129 Bologna, Italy

L. Venturi
Department of Physics and Astronomy, University of Bologna,
Via Irnerio 46, 40126 Bologna, Italy

V. Morandi and L. Ottaviano (eds.), *GraphITA*, Carbon Nanostructures,
DOI 10.1007/978-3-319-58134-7_14

[3]. These exhibited excellent electrical and thermal conductivity, high surface area and good structural integrity.

Since their first development there have been many investigations into applications of graphene foams. Without further modification they have demonstrated great potential in such diverse applications as high sensitivity detection of NH_3 and NO_2 gas [4], supercapacitors [5], and conductive scaffolds for the proliferation of neural stem cells [6]. In addition, because of their porous nature, composites with graphene foams may easily be produced by filling the pores with other materials. For example, their capacitive behaviour has been further improved by decoration with materials such as cobalt oxide nanobundles [7], zinc oxide nanorods [8], or carbon nanotubes [9]. With the addition of copper oxide nanoflowers they have been used as free-standing biosensing platforms [10], and with tungsten oxide nanoparticles, promising photocatalytic materials [11]. Composites with polymers have also been investigated, for example with Teflon to produce superhydrophobic coatings [12], or with PDMS to create stretchable strain sensors [13]. The field continues to grow rapidly, with new applications constantly being proposed.

However, the graphene structures grown by this technique are limited by the structure of the commercially available templates, which typically have pore sizes in the range of 200–400 μm. In this case most of the volume is occupied by void space rather than the functional graphene material, which is a disadvantage in applications such as energy storage, where a high volumetric energy density is required. Thus a smaller pore size is often desirable. To achieve this, in 2014 Goh et al. infiltrated a Ni foam with Ni particles in a slurry with PMMA (which acted as both support for the Ni particles and solid carbon fuel source), then performed low pressure CVD and template etching processes to produce a graphene foam whose voids were filled with graphene "eggshells" [14]. This resulted in a Li-ion battery anode material with greatly improved charge transport, ion transport and rate of lithium intercalation, as well as enhanced volumetric density, electrochemical activity and mechanical stability.

In this work we describe the development of an alternative procedure for producing high-density graphene foams (which we call HD-GF), without the need for a commercial Ni foam backbone or a solid carbon fuel source and under atmospheric pressure conditions. Instead, simply by heating Ni nanoparticles we can create a dense interconnected 3D network on which graphene can be grown. In this way we synthesised HD-GF with a pore size down to 1 μm, which have excellent prospectives for applications that require a high surface area to volume ratio.

2 Experimental

2.1 *Preparation of Low-Density Graphene Foams (LD-GF)*

Low-density graphene foams (LD-GF) were grown on commercially available nickel (Ni) or copper (Cu) foam templates (Graphene Supermarket, 320 gm^{-2} areal density and thickness 1.2 mm) by Chemical Vapour Deposition in a hot-wall tube furnace (diameter 5 cm) using a methane (CH_4) precursor. The templates were washed by ultrasonication in dilute hydrochloric acid, then deionised water, and finally acetone before being placed inside the furnace and heated to 1000 °C under a hydrogen (H_2) flow of 50 sccm. After annealing under these conditions for 30 min, the flow of H_2 was increased and a flow of CH_4 was introduced (H_2: CH_4 = 500:50 sccm). Following a 10 min deposition time, the sample was removed from the furnace and allowed to cool in flowing argon or nitrogen for at least 2 h. The LD-GF grown on Cu foam templates were drop-coated with PMMA to support the more fragile structure during subsequent processing. To remove the sacrificial Ni and Cu templates, the samples were immersed overnight either in 17 vol.% nitric acid (HNO_3) or in 4.5 vol.% iron (III) chloride ($FeCl_3$) followed by 10% hydrochloric acid (HCl) heated to 80 °C. Finally, they were thoroughly rinsed in deionised (DI) water then allowed to dry completely before further handling. In the case of LD-GF grown on Cu foams, the PMMA was removed by immersion in cold acetone for at least 2 h.

2.2 Preparation of High-Density Graphene Foams (HD-GF)

High-density graphene foams (HD-GF) were grown on Ni nanoparticles (Sigma-Aldrich, <100 nm average particle size), or a combination of Ni and nickel (II) oxide nanoparticles (NiO, Sigma-Aldrich, <50 nm particle size), both as loose powders and mechanically pressed pellets (diameter 1.2 cm). The samples were first heated to 600 °C and held at this temperature for 10 min in a H_2 atmosphere to partially melt the surface of the nanoparticles and sinter them together into a network. The temperature was then increased to 1000 °C and graphene was grown under the same conditions as those described for LD-GF. NiO was reduced to Ni during the annealing step. The sacrificial template was removed as before, but a longer period (several days) of immersion in the etching solution were required before no more colour change was observed, indicating that all the Ni had been removed.

2.3 Characterisation

Scanning Electron Microscope (SEM) observations were performed with a ZEISS LEO 1530 microscope equipped with a Schottky emitter, operated at 5 keV and collecting secondary electrons by means of an In-Lens detector. Transmission Electron Microscope (TEM) observations were performed on samples ultrasonicated in IPA and drop-cast on a holey carbon grid, using a FEI Tecnai F20 ST microscope operated at 120 keV. The images were captured with a Gatan MSC794 camera.

Raman spectroscopy was performed using a Renishaw 1000 instrument with a source laser wavelength of 632.8 nm, lateral spatial resolution of 1 μm and spectral resolution of 2 cm^{-1}. X-Ray Diffraction (XRD) was performed using a SmartLab-Rigaku diffractometer, equipped with a rotating anode (Cu Kα $\lambda = 1.54180$ Å) followed by a parabolic mirror to collimate the incident beam, and a series of variable slits (placed before and after the sample position).

3 Results and Discussion

3.1 *Optimisation of Low-Density Graphene Foams*

The choice of either Ni or Cu foam for the substrate has a significant effect on the properties of the resulting LD-GF. Using a Ni foam template and precursor flow rates of 50 sccm CH_4 and 500 sccm H_2, a continuous, free-standing LD-GF was produced (Fig. 1b, c). From SEM observations on >10 locations, the wall thickness was estimated to be 10–20 nm. Characteristic wrinkles were observed in the foam, arising from the mismatch in thermal expansion coefficients between Ni and graphene [15]. The interior appeared hollow, indicating that Ni etching was successful. However, XRD measurements on LD-GF etched with HNO_3 showed residual Ni (Fig. 1g), which most likely remained due to the gas bubbles produced by the reaction of Ni with HNO_3. These gas bubbles could block the infiltration of etchant to some channels of the internal structure. To overcome this, etching was performed instead in $FeCl_3$, which has no gaseous products on reaction with Ni, only soluble salts. These were washed away in hot HCl, which also served to complete the etching process. XRD measurements confirmed that in this way, the resulting LD-GF had few traces of Ni when $FeCl_3$ was used at room temperature (Fig. 1g), and no traces of Ni when it was at 80 °C. However, in this case there was some residual Fe observed (Fig. 1g).

TEM diffraction measurements confirmed the polycrystalline graphite nature of the LD-GF, although the thickness of the walls was too large to observe the individual layers of graphene (Fig. 1e, f). The quality of the graphene was further investigated by Raman spectroscopy (Fig. 1h). The lack of the D peak (at approximately 1340 cm^{-1}) in the spectrum showed that the LD-GF had minimal defects. The 2D peak was broad, consisting of two obvious components, and the ratio of peak intensities (I_{2D}/I_C) varied between 0.3 and 0.6, once again confirming that the LD-GF consisted of multiple layers of graphene, approaching graphite [16]. It is worthwhile noting that thinner LD-GF featuring fewer layers of graphene may also easily be grown by varying parameters such as the amount of CH_4 precursor, the growth time and/or temperature and the cooling rate. However, thinner foams

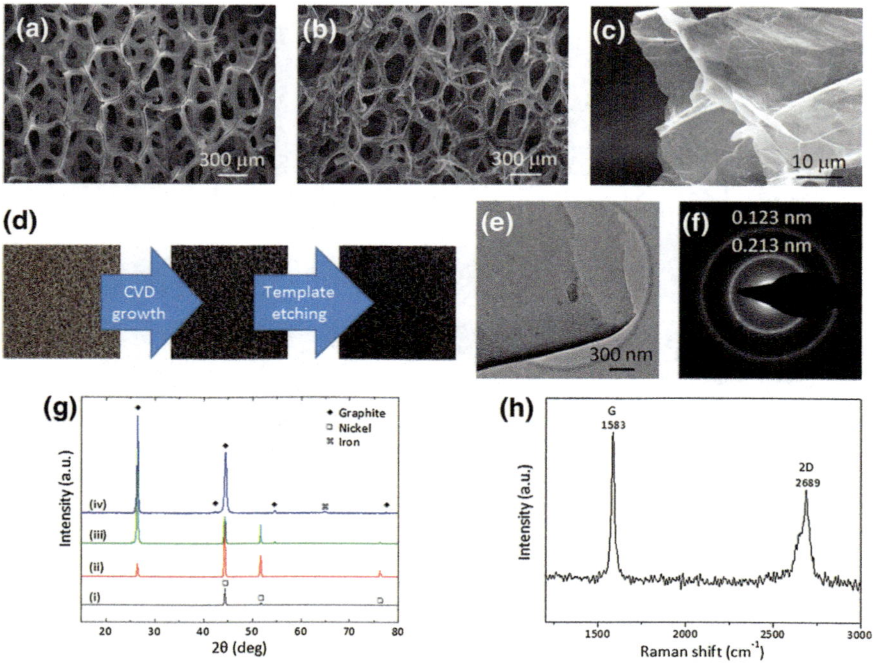

Fig. 1 Characterisation of LD-GF from Ni foam precursor. SEM images of **a** nickel foam as received, **b**, **c** LD-GF. **d** Photographs of foam at different stages of the synthesis process. **e** TEM image of LD-GF. **f** Electron diffraction pattern from the foam area in (**e**). **g** XRD patterns of (i) Ni foam, LD-GF etched with (ii) HNO_3, (iii) $FeCl_3$ followed by hot HCl and (iv) hot $FeCl_3$ followed by hot HCl. **h** Raman spectrum of LD-GF

are less robust and require a PMMA support layer during the etching process so that they do not collapse [3].

Using the same precursor flow rates on a Cu foam template, no free-standing graphene foam could be recovered, even using PMMA as a support during Cu etching. When the PMMA was removed, the foam completely disintegrated. After increasing the concentration of CH_4 (H_2:CH_4 = 200:100 sccm) some material could be collected on a silicon wafer substrate, but it was completely collapsed (Fig. 2b). TEM imaging revealed thicknesses ranging from 1 to 20 graphene layers based on the lattice fringes observed (Fig. 2c). This was confirmed by the Raman spectra, in which the I_{2D}/I_G ratio of 1.2 and the broad but single-component 2D peak indicated on average few layer graphene (Fig. 2f). In the Raman spectrum, there also appeared a strong D peak indicating that the graphene had many defects and was consequently of lower quality than the equivalent LD-GF from Ni foam. Increasing the precursor concentration even further (H_2:CH_4 = 100:100 sccm) finally led to a free-standing foam, although it had many cracks (Fig. 2d, e) and even though the wall thickness was estimated to be approximately 60–70 nm by SEM, it tended to fragment when handled after the PMMA support was removed.

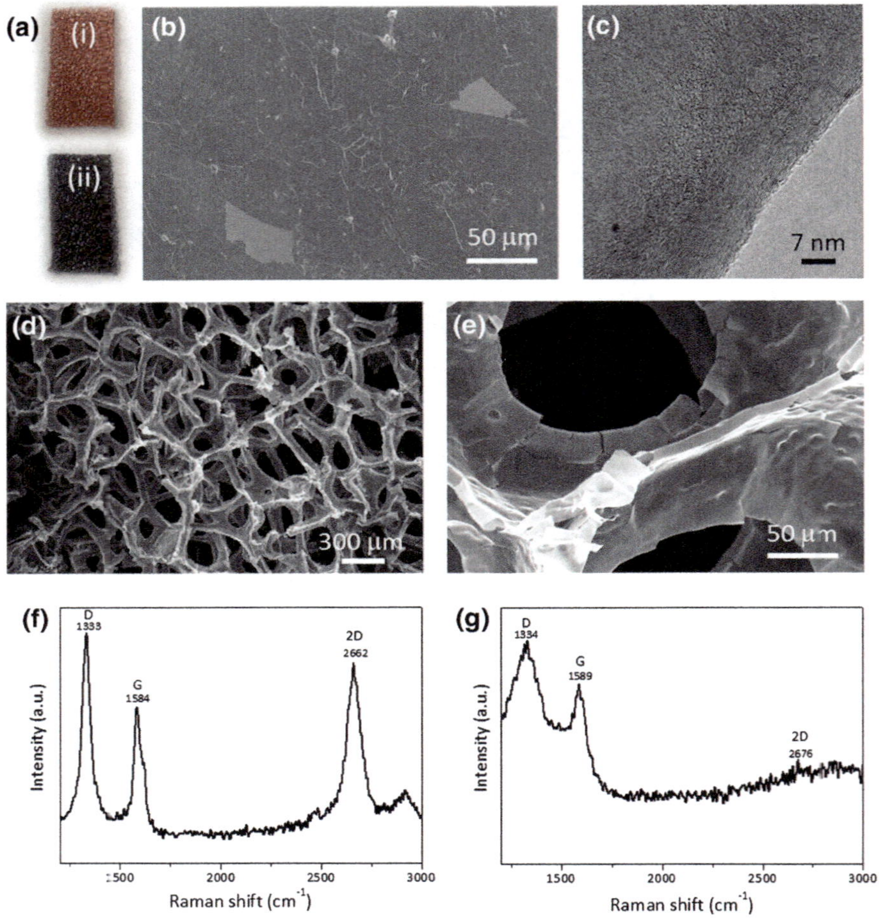

Fig. 2 Characterisation of LD-GF from Cu foam precursor. **a** Photographs of Cu foam + graphene (i) and LD-GF + PMMA (ii). **b** SEM image of LD-GF from H_2:CH_4 = 200:100 sccm precursor concentrations. **c** TEM image of the same foam as in (**b**). **d**, **e** SEM images of LD-GFs from H_2:CH_4 = 100:100 sccm at different magnifications. **f** Raman spectrum of LD-GF grown from H_2:CH_4 = 200:100 sccm. **g** Raman spectrum of LD-GF grown from H_2:CH_4 = 100:100 sccm

A very large D peak in the spectrum of Fig. 2g evidences the presence of many defects. However, virtually no 2D peak is visible and the spectrum is similar to that of an amorphous carbon film or of a fine grain polycrystalline multilayered structure.

The difference in the mechanism for graphene growth on Ni and Cu is well documented and is related to the different solubility of carbon (C) in each metal [17]. Because of the high solubility of C in Ni, the growth of graphene occurs by a dissolution-surface segregation-precipitation process, which leads directly to a uniform, continuous, usually multilayer coating of graphene. On the other hand, because of the low solubility of C in Cu, graphene is grown on Cu by surface adsorption of C followed by its migration, which generally leads to a monolayer of graphene at low precursor concentrations: this is not strong enough to create a free-standing foam. At high precursor concentrations, it appears that in a tortuous structure like a Cu foam, areas of thicker and thinner graphene are grown depending on where C has been able to adsorb and migrate, creating a highly defected coating with many cracks. Consequently, only Ni was chosen for the development of HD-GF.

3.2 *Development of High-Density Graphene Foams*

Building on what was learnt from synthesising LD-GF, a procedure was developed for the synthesis of HD-GF. Templates were first produced by simply placing a heap of Ni nanoparticles on a quartz sample holder and heating it to 600 °C to partially melt their surfaces [18] and cause them to fuse together into a 3D network. This has the advantage of being easily scalable since the size of the resulting foam is determined only by the amount of nanoparticles that are used. However, because there is no control over the position of the particles with respect to each other, there is also poor control over the final pore-size distribution (Fig. 4b), and both micro- ($\sim$1 μm) and macro- (>10 μm) pores were observed, as evident in the SEM image of the resulting HD-GF (HD-GF-1, Fig. 3a). These are, however, considerably smaller on average than those of the LD-GF grown on commercially available Ni foams (50–400 μm, Fig. 4a). A wall thickness of approximately 10 nm was estimated from SEM measurements. TEM imaging of various fragments revealed a variety of wall thicknesses, anywhere from 5 to more than 100 graphene layers based on the lattice fringes observed (Fig. 3c, d). Given that the CH_4 precursor must penetrate through the dense network of nanoparticles within a relatively short deposition time (10 min), it is unsurprising that different wall thicknesses are observed, based on the position within the foam. Electron diffraction measurements confirmed that the graphene is polycrystalline (Fig. 3e).

HD-GF-1 was quite fragile, most likely due to the lack of uniformity between light areas with large voids and heavier areas with many interconnecting branches, causing it to fragment when not handled carefully. However, to ensure that this was not due to the nature of the graphene grown, the quality of the graphene was analysed by Raman spectroscopy. Like the LD-GF produced on a Ni foam template, no D peak was evident, indicating that there were few defects present (Fig. 3b). The 2D peak was relatively broad and an I_{2D}/I_G ratio of approximately 0.5 was observed, indicating multilayer graphene, in agreement with the SEM and TEM measurements.

In order to better control the pore-size distribution and produce a more uniform and robust HD-GF, before heating the nanoparticles were mechanically pressed to

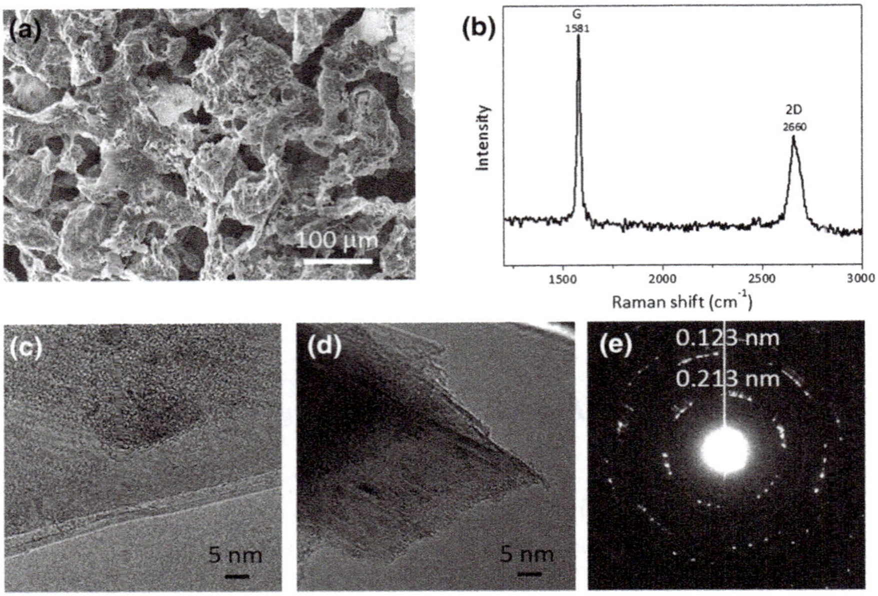

Fig. 3 Characterisation of HD-GF-1. **a** SEM image of the surface of the HD-GF. **b** Raman spectrum acquired from the sample in (**a**). **c**, **d** TEM images of, respectively, a few layer region and a many layer region. **e** Electron diffraction pattern from the area of the sample in (**c**)

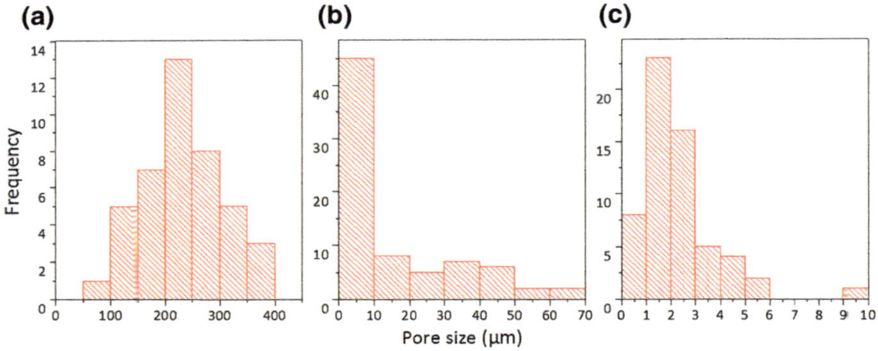

Fig. 4 Pore-size distributions based on a series of SEM images. **a** LD-GF from commercial Ni foam template. **b** HD-GF-1 from a heap of Ni nanoparticles. **c** HD-GF-2 from a pellet of Ni/NiO nanoparticles pressed under 50 kg/cm^2

form pellets using a 1.2 cm die set. It was expected that varying the amount of force used would lead to different pore-size ranges. However, using even the minimum pressure reachable using a hydraulic press (440 kg/cm^2), the resulting pellets were too compact, leading to predominantly sheets of graphene with only small areas of

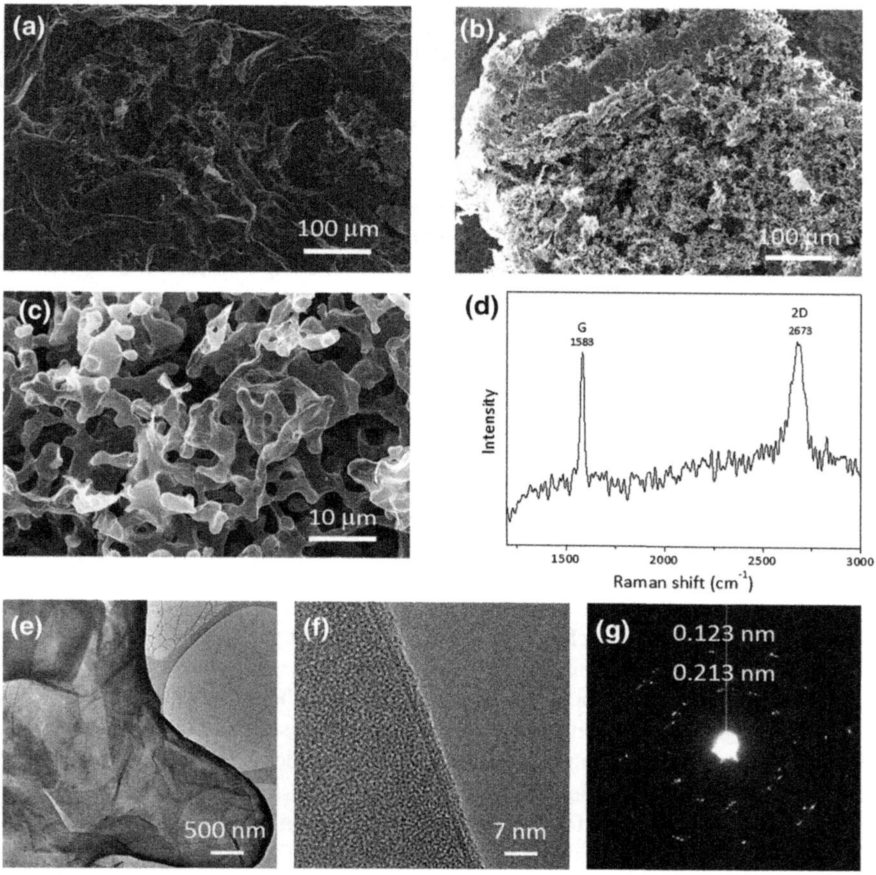

Fig. 5 **a** SEM image of GF from pellet of Ni nanoparticles pressed under 440 kg/cm^2. **b**, **c** SEM images of the HD-GF-2. **d** Raman spectrum from HD-GF-2 sample. **e**, **f** TEM images of the HD-GF-2 foam. **g** Electron diffraction pattern from the area in (**f**)

porosity where the nanoparticles had resisted the tight packing (Fig. 5a). Seemingly the particles were so close to each other that the graphene grown on top was able to join together as it precipitated rather than wrapping around each nanoparticle (or aggregate of nanoparticles). To produce a less compact pellet, the die was pressed by hand. In this way, a pressure of approximately 50 kg/cm^2 could be achieved, but this was insufficient to hold the Ni nanoparticles together, causing the pellet to fall apart as soon as it was removed from the die set. To solve this problem, a small amount (15–40 mass%) of NiO nanoparticles was introduced, which acted to bind the nanoparticles together so that a pellet could be made using this lower pressure. The ability of the NiO nanoparticles to stick together more strongly than Ni nanoparticles could be related to the presence of oxygen, and in particular its non-stoichiometric ratio with Ni, which is evident from the colour of the particles:

black rather than green [19]. This leads to coordinatively unsaturated sites at the surface of the particles, so they more readily interact with their surroundings and bind more strongly to each other. They are also susceptible to the absorption of gases from the atmosphere, which could explain why the amount of NiO required to hold the pellets together was variable, possibly based on humidity level. Despite the presence of NiO, the mechanism for the subsequent growth of graphene does not change since during the annealing process in a high-temperature hydrogen atmosphere NiO is reduced to Ni, leading to the same behaviour as for Ni nanoparticles alone. Indeed, similar HD-GF were synthesised using templates of only NiO nanoparticles, but these shrunk considerably during the annealing process due to the removal of oxygen, leading to very small foam samples that were not practical for further experimentation. The final amount of NiO used was therefore the minimum required to hold the pellets together so that the final foam was approximately the same size or only slightly smaller than the original pellet.

The HD-GF produced using the new template, a pellet of a combination of Ni and NiO nanoparticles, had a significantly more controlled particle size distribution than HD-GF-1, as evident from the SEM images (HD-GF-2, Fig. 5b, c). On the top and bottom of HD-GF-2, there was a layer of graphene sparsely scattered with holes of around 1 μm which corresponds to where the nanoparticles were compacted most strongly in the template when it was pressed. However, beneath this layer a network of interconnected hollow branches of graphene were observed. The pores were mostly in the range of 1–10 μm, around two orders of magnitude less than the LD-GF (Fig. 4c). Observations were made on many different parts of the foam and the aspect was similar throughout, indicating good homogeneity. As estimated from the SEM images, the thickness of graphene was less than 10 nm. Indeed, by TEM, around 10–30 layers of graphene could be observed (Fig. 5e, f) and the polycrystalline graphite nature was confirmed as before by diffraction measurements (Fig. 5g).

To analyse the quality of the graphene in HD-GF-2, Raman spectroscopy was once again employed (Fig. 5d). The HD-GF-2 sample was first fragmented slightly to ensure that measurements were taken from the inside of the foam in the 3D network region rather than from the top or bottom graphene layers. As for LD-GF on Ni and HD-GF-1, no D peak was visible in the Raman spectrum, indicating minimal defects. Additionally, the I_{2D}/I_G ratio was higher than before (approximately 1), indicating fewer graphene layers, although the 2D peak was still broad. The reduction in graphene deposited could be due to the denser nanoparticle network, which reduced the percolation of CH_4 precursor through the material and resulted in a comparatively lower dissolution of C.

4 Conclusion

We first explored the synthesis of graphene foams from commercial metal foam templates (LD-GF) to find the optimal growth and processing conditions. LD-GF grown from Ni templates were high quality and robust, while those from Cu templates had many defects and tended to fragment. Unlike previous studies, a polymer support layer was not required during the removal of the Ni template by chemical etching. The most effective etchant was found to be hot $FeCl_3$ followed by HCl. Using this information, we developed a procedure for the synthesis of graphene foams with a higher pore density (HD-GF). By using a template of Ni and NiO nanoparticles pressed lightly into a pellet and annealed to form a 3D network, we produced a high quality HD-GF with the majority of pores in the range of 1–5 μm, significantly smaller than the 50–400 μm achievable with commercially available foam templates. These have the potential for application in areas where a larger amount of the active graphene material needs to be available in a given volume, for example in energy storage or catalysis.

Acknowledgements We acknowledge funding by the European Union Seventh Framework Programme under Grant Agreement No. 604391 Graphene Flagship.

References

1. Dikin, D.A., et al.: Preparation and characterization of graphene oxide paper. Nature **448**, 457–460 (2007)
2. Xu, Y., Sheng, K., Li, C., Shi, G.: Self-assembled graphene hydrogel via a one-step hydrothermal process. ACS Nano **4**, 4324–4330 (2010)
3. Chen, Z., et al.: Three-dimensional flexible and conductive interconnected graphene networks grown by chemical vapour deposition. Nat. Mater. **10**, 424–428 (2011)
4. Yavari, F., et al.: High sensitivity gas detection using a macroscopic three-dimensional graphene foam network. Sci. Rep. 1 (2011)
5. Chen, W., Fan, Z., Zeng, G., Lai, Z.: Layer-dependent supercapacitance of graphene films grown by chemical vapor deposition on nickel foam. J. Power Sour. **225**, 251–256 (2013)
6. Li, N., et al.: Three-dimensional graphene foam as a biocompatible and conductive scaffold for neural stem cells. Sci. Rep. **3** (2013)
7. Deng, W., Sun, Y., Su, Q., Xie, E., Lan, W.: Porous CoO nanobundles composited with 3D graphene foams for supercapacitors electrodes. Mater. Lett. **137**, 124–127 (2014)
8. Li, X., Wang, Z., Qiu, Y., Pan, Q., Hu, P.: 3D graphene/ZnO nanorods composite networks as supercapacitor electrodes. J. Alloys Compd. **620**, 31–37 (2015)
9. Wang, W., et al.: Three dimensional few layer graphene and carbon nanotube foam architectures for high fidelity supercapacitors. Nano Energy **2**, 294–303 (2013)
10. Ma, Y., et al.: 3D graphene foams decorated by CuO nanoflowers for ultrasensitive ascorbic acid detection. Biosens. Bioelectron. **59**, 384–388 (2014)
11. Azimirad, R., Safa, S.: Preparation of three dimensional graphene foam–WO3 nanocomposite with enhanced visible light photocatalytic activity. Mater. Chem. Phys. (2015). doi:10.1016/j.matchemphys.2015.06.043
12. Singh, E., et al.: Superhydrophobic graphene foams. Small **9**, 75–80 (2013)
13. Jeong, Y.R., et al.: Highly stretchable and sensitive strain sensors using fragmentized graphene foam. Adv. Funct. Mater. n/a–n/a (2015). doi:10.1002/adfm.201501000
14. Goh, B.-M., et al.: Filling the voids of graphene foam with graphene 'Eggshell' for improved lithium-ion storage. ACS Appl. Mater. Interfaces. **6**, 9835–9841 (2014)
15. Wang, L., Li, X., Guo, T., Yan, X., Tay, B.K.: Three-dimensional Ni(OH)2 nanoflakes/graphene/nickel foam electrode with high rate capability for supercapacitor applications. Int. J. Hydrogen Energy **39**, 7876–7884 (2014)

16. Ferrari, A.C., et al.: Raman spectrum of graphene and graphene layers. Phys. Rev. Lett. **97** (2006)
17. Li, X., Cai, W., Colombo, L., Ruoff, R.S.: Evolution of graphene growth on Ni and Cu by carbon isotope labeling. Nano Lett. **9**, 4268–4272 (2009)
18. Matsuno, M., et al.: In situ sintering of Ni nanoparticles by controlled heating. Microsc. Microanal. **17**, 524–525 (2011)
19. Gonzalez-Elipe, A.R., Holgado, J.P., Alvarez, R., Munuera, G.: Use of factor analysis and XPS to study defective nickel oxide. J. Phys. Chem. **96**, 3080–3086 (1992)

Protein-Based Nanostructures and Their Self-assembly with Graphene Oxide

R. Ippoliti, M. Ardini, L. Di Leandro, F. Giansanti, A. Cimini, L. Ottaviano, V. Morandi, L. Ortolani and F. Angelucci

Abstract Proteins are hetero-polymers made-up of single building blocks (aminoacids) whose composition determines folding and final architecture. Some proteins are able to undergo self-assembly process enabling the formation of ordered molecular aggregates that in some cases assume conformations particularly suitable to nanotechnological applications. In this work we describe the properties of a ring-like decameric protein, Peroxiredoxin (Prx), to build composite materials interacting with or catalyzing the formation of selectively metal nanoparticles that can be trapped over the surface of nanostructured graphene oxide (GO) sheets. We demonstrate furthermore the ability of Prx to guide the formation of 3D layers of GO embedding metal nanoparticles in the composite material. These composites are discussed as possible precursors to electronic and chemical devices.

During the last 15 years bionanotechnology emerged as a new effective discipline where physical and biological principles are perfectly combined to generate new functional materials and devices of scientific and industrial interest [21, 29]. Bionanotechnology, in fact, introduced new approaches for challenging all endeavors related to downscaling the size of building block materials to be used for manufacturing new smart nanoscale devices, in order to further broaden the range of technological applications. Moreover, such nanodevices are required to hold special functionalities while keeping enough flexibility to integrate into a large number of different environments [9, 16]. It is widely recognized that the self-assembly or bottom-up approach is the simplest and straightforward strategy to access the

R. Ippoliti (✉) · M. Ardini · L. Di Leandro · F. Giansanti · A. Cimini · F. Angelucci
Department of Life, Health and Environmental Sciences,
University of L'Aquila, 67100 L'Aquila, Italy
e-mail: Rodolfo.ippoliti@univaq.it

L. Ottaviano
Department of Physics and Chemistry, University of L'Aquila, L'Aquila, Italy

V. Morandi · L. Ortolani
National Research Council, Institute for Microelectronics and Microsystems, Bologna, Italy

V. Morandi and L. Ottaviano (eds.), *GraphITA*, Carbon Nanostructures,
DOI 10.1007/978-3-319-58134-7_15

manipulation of complex nanostructured materials [24, 25, 33, 34], difficult to synthesize through traditional methods, and organize them into large-scale assemblies with smart functionalities. This approach should ensure that those functionalities are well integrated into the single nanomaterial and working efficiently without interfering each other such that the final device would operate with the desired response [9].

The bionanotechnological approach is directly inspired by nature where a huge set of "living" nanodevices are ready to use as a consequence of millenniums of evolution. This concept was foreseen in late 50s when Feynman speculated the use of such biological nanometric machines, i.e, those making cells active, moving, manufacturing and wiggling, for technological applications due to their ability to self-assemble and convert chemical forces in a repetitious fashion thus obtaining all kinds of weird effects [12]. When operating in the nanoscale realm, primarily in the challenging "1–100 nm niche", nanotechnology offers a wide range of tools and applications. Here, biomolecules such as DNA, lipids, proteins and peptides perfectly fit with these dimensions thus providing a first obvious advantage. Furthermore, the size of their nanoscale structures also allows them to interact with living matter, e.g, with the cell surface or within the cell, often without altering the biochemical and biological properties of the reactants [15, 22].

Even though peptides and DNA have been widely considered as potential building blocks or tectons, they lack the plethora of functionalities typical of proteins. Proteins, in fact, offer enormous diversity in three-dimensional (3D) architecture and virtually endless possibilities for modification and functionalization so it comes as with no surprise that protein nanotechnology has become a very active field capturing both the excitement of recent advances in nanotechnology and harnessing our exponentially expanding knowledge of protein science [13]. However, in contrast to DNA and peptides, the rational design of protein nanostructures is much more complicated due to the complex interactions involved in the native protein fold. As a consequence, the structural folding of most natural proteins cannot be easily predicted from their primary structure and therefore, de novo design of completely new protein nanostructures is hardly to be achieved, especially in the absence of existing native folds available for structural comparisons. Nevertheless, the design of protein assemblies is going beyond the proof of principles thanks to new approaches that take advantage on the symmetric properties of natural existing protein domains. These domains are linked both by genetic manipulation or by chemical cross-linking in order to induce their self-assembly, leading to the formation of symmetric intermolecular protein nano-assemblies, as lattices and cage-like structures [14]. However, proteins already exist in nature as symmetric monodisperse nano-assemblies like cages, rings and tubes all offering useful starting points for further modification which can be engineered to produce semi-artificial structures realistically looking as promising basic components for future work. In addition, protein systems offer a number of other advantages such as synthesis under ambient conditions with no toxic by-products. Finally, proteins offer many other advantages over non-biological materials. For example, they may be able to act at the interface between inorganic and biological systems, e.g, carbon material and cells, beyond the ease of production in large (industrial) scale from cheap bacterial expression systems [16]. Yet, the high specificity and strength of the antibody–antigen interaction is suitable to control the localization of the building block during the construction of a complex nano-architecture, while enzymes are ready-to-use nanoreactors where specific reactions could be accomplished or templated for the confined synthesis of inorganic nanostructures such as metals and semiconductors [9].

Despite the advantages of the self-assembling protein tectons, limitations come out when directly applying them for practical purposes. For example, while some proteins exhibit remarkable physical characteristics, e.g, the tensile strength of spider silk which is the strongest biological fibers known, they cannot compete with those of non-biological materials, e.g, the tensile strength of carbon nanotubes is at least an order of magnitude higher. Furthermore, the use of carbon nanotubes and other related materials in electronics like semiconductors or superconductors is likely affordable, something that biological molecules are unlikely to be capable of [16]. In this context, the use of carbon-based materials combined to biomolecules has been taken into the limelight in the last few years generating an exciting and pretty new research area. Here, the extraordinary properties of graphene and its related products, i.e, graphene oxide (GO) and reduced graphene oxide (rGO), are now meant to be top players in that concerning biomolecule-mediated self-assembling of new active materials. For this purpose, the two-dimensional (2D) structure of graphene is converted into stable and accessible 3D architectures without impairing its properties. Indeed, graphene undergoes self-stacking leading large suppression of conductivity, mechanical strength, and accessible surface area with respect to the individual layers. Therefore, tailoring the desired geometries of 3D graphene at nano/microscale has a great impact on the material's performances [37]. Furthermore, it is now recognized that discerning such lamellar architecture in terms of number of layers has become mandatory to further classify graphene nanostructures due to the explosive revolution of the carbon world [32].

Reaching a 3D self-assembly of graphene typically starts using GO as a precursor. GO exhibits flexibility of the lattice, allowing it to contort and conform to non-planar architecture either in single or stacked configuration [32]. A force balance between the van der Waals attractions from its carbon plane and the electrostatic repulsions from the edges makes GO an amphiphilic molecule in aqueous solutions and near-neutral pH. Controlled breaking of such balance induces GO layers to stack into hierarchical 3D composites found as foams, sponges, and aerogels [11, 20, 23]. Biological macromolecules, for example, are now confirmed as smart cross-linking compounds for GO [39]. The first case has been reported in 2010 where the natural, high abundant single-stranded DNA was left reacting with the planar smooth surface of GO layers through non-covalent π–π interactions. This strategy led to lamellar bio-nanocomposites with intercalated DNA-cytochrome c mixtures that could be accessed by small molecules such as sodium dithionite [26].

This scenario showed a brand new way path for obtaining 3D graphene-based materials with tailor-made properties and improved biocompatibility directly coming out from the biological moiety itself. In this context, proteins could really play a prominent role. First, proteins can be immobilized on GO layers in aqueous solution forming stable bio-nanocomposites as demonstrated for serum proteins [28, 31] This is accomplished without using any cross-linking reagent but simply exploiting non-covalent interactions, namely hydrogen, van der Waals and π–π bonds. Second, due to its flat large surface, GO could provide a molecular platform for multiple protein components making more feasible the study of the interaction mechanisms through classical techniques, e.g, AFM and TEM [38]. Third, it is also reported that GO-immobilized protein enzymes show improved activity and structural stability thus favoring their potential use in practical applications that can be found as sophisticated nano-machines endowed with multiple structural configurations and enzymatic properties which are very hard to achieve through chemical synthesis [13].

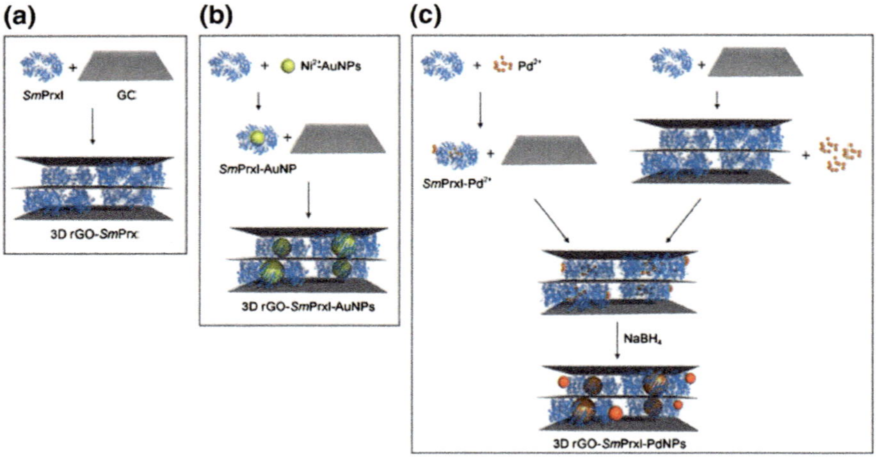

Fig. 1 **a** The *Sm*PrxI rings adhere flat over the smooth planar surface of single GO layers and partially reduce them to induce self-assembly of a 3D multi-layer rGO-*Sm*PrxI hybrid composite. **b** The same approach is exploited to obtain a rGO-*Sm*PrxI-AuNPs composite. In this case, the protein is pre-conjugated with Ni^{2+}-functionalized AuNPs through the interaction between the N-terminal histidine-rich binding sites and the Ni^{2+} ions. **c** Likewise, the same interaction, together with the capability of protein surfaces to capture metal ions, is exploited to stably trap Pd^{2+} ions onto the ring template prior to chemical reduction towards a rGO-*Sm*PrxI-PdNPs material (reproduced from Ardini et al. [4])

Taking advantage of such inspiring though limited information, we recently explored the interaction between GO and the ring-like protein peroxiredoxin (Prx), gaining new insights into the possibility of using it to build-up ultralight 3D multi-layer composites of rGO doped with metal NPs [3, 4]. This approach demonstrated that a unique self-assembling proteinaceous reagent with a discrete architecture, i.e, the ring-like Prx, is able to (i) induce ordered stacking of single GO layers, (ii) partially reduce them into rGO and (iii) dope the resulting 3D material with metal NPs directly attached onto the Prx ring template or de novo synthesized inside it (Fig. 1, reproduced from Ardini et al. [4]).

Prx appears promising for nanotechnological purposes as it naturally undergoes hierarchical self-assembly to form discrete nanostructures such as decamer rings and stacks [1, 2, 27]. Another fundamental feature of the Prx ring complex is the internal cavity which is 6 nm wide, a value larger than other similar protein rings such as Hcp1 (4 nm), SP1 (2.5 nm), TRAP (2 nm), GroEL (4.5 nm), and thermosome (5 nm). Furthermore, all N-terminal regions of the ring complex point towards the cavity while the C-terminal ones are exposed outside, forming two independent potential binding sites for ligands if appropriately engineered. Finally, Prx exhibits an uncommon long-term stability against heating treatments without significant precipitation and impairment of the ring architecture. First, we were able to induce and govern the formation of Prx nanotubes and exploit such structures as hollow scaffolds to align ultrasmall AuNPs into 1D peapod-like nanostructures. To accomplish this task, Ni^{2+}-functionalized AuNPs were incorporated inside the Prx ring cavity whose N-terminal regions were modified with histidine-rich metal-binding sequences (Fig. 2).

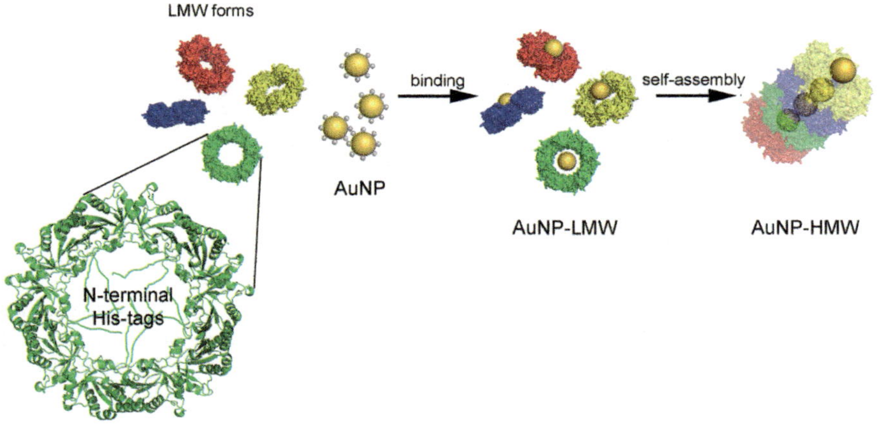

Fig. 2 Nickel-functionalized gold nanoparticles (AuNPs) are encapsulated within the ring cavity yielding hybrid adducts with torusphere architecture (AuNP-LMW adducts). These complexes can stack in turn to form 1D peapod-like tubular chains (AuNP-HMW adducts). Modified from Ardini et al. [3]

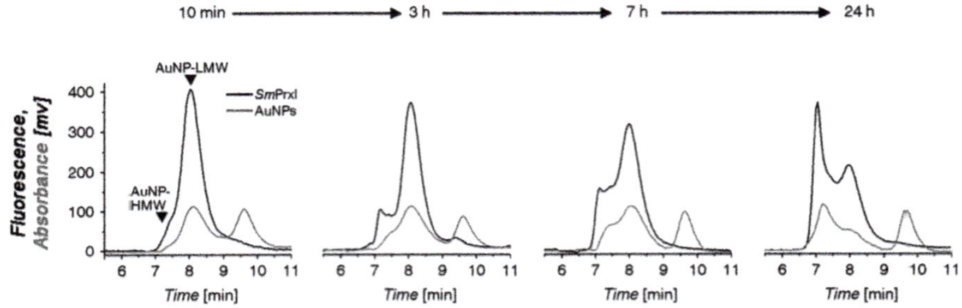

Fig. 3 Time-course formation of the gold-protein adducts. The protein constantly co-elutes with the AuNPs indicating a lasting interaction upon mixing. Both the LMW and HMW species are present and interact with the nanoparticles yielding to the AuNP-LMW and AuNP-HMW adducts, respectively. As the reaction goes on, a slow aggregation process clearly occurs with the time favoring the formation of the AuNP-HMW adducts to the detriment of the AuNP-LMW ones (reproduced from Ardini et al. [3])

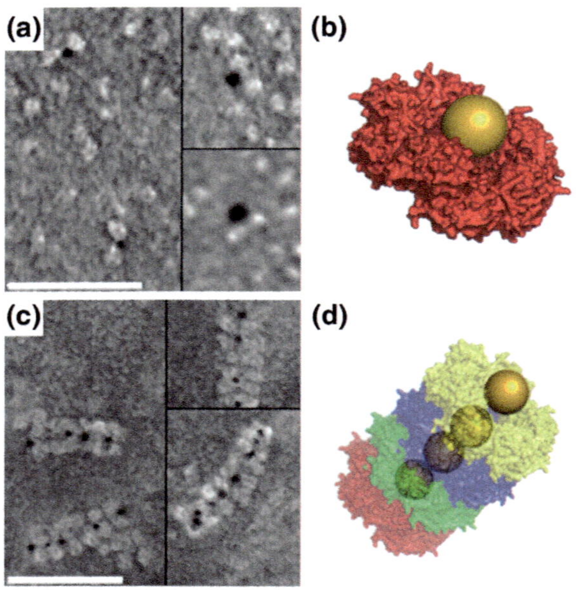

Fig. 4 Structural and topographical characterization of the gold-Prx adducts. **a** TEM imaging of the AuNP-LMW hybrids shows discrete torusphere-like particles made of single large gold nanoparticles linked to single protein rings. **b** 3D molecular model of AuNP-LMW torusphere. **c** The AuNP-HMW adducts are imaged by TEM as peapod-like complexes made of repetitive AuNP-LMW chains containing nanoparticles. **d** 3D molecular model of the AuNP-HMW nano-peapods. Scale bars correspond to 40 nm (reproduced from Ardini et al. [3])

This assembly mechanism accounts for two subsequent and synergistic steps. First, the binding of AuNPs to the single rings to form single Prx-AuNPs adducts. The second step is a direct consequence of the proximity between such adducts as once getting they are able to stack directionally through both the self-recognition properties of the ring's complementary surfaces and the ability of the AuNPs to join the rings as showed by GFC experiments (Fig. 3).

The process is likely to be mediated by several coordination bonds between the Ni^{2+} ions tethered on the NPs' surface and the N-terminal His-tails within the protein cavity. This hypothesis is supported by several reasons as follows. The engineered N-terminal regions of the Prx ring contain six histidine and four aspartate residues which are often found in metal-binding proteins [35], resulting in the presence of 100 potential binding sites making a 140 nm^3 environment suitable for stably trapping metal ions. Furthermore, TEM images of the gold-protein adducts showed that the Ni^{2+}-functionalized AuNPs co-localize with the N-terminal His-tails (Fig. 4).

Thus, it is likely that the metal ions act as polymerizing agents and increase the local concentration of the gold-protein conjugates inducing the transition from single rings to stacked nanotubes with aligned AuNPs chains within the protein shell.

According with these findings, 3D rGO-based composites have been assembled and functionalized with metal NPs, by adsorbing the protein rings on single 2D GO layers. To the best of our knowledge this represents a unique "green" synthetic route to achieve assembly, chemical reduction and functionalization with metal NPs of graphene composites simply by using a single "multitasking" protein reagent. Building-up 3D graphene nanostructures enriched with metal nanoparticles is a very active research area as the resulting materials exhibit synergistic properties like new "nanoreactors" with enhanced chemical–physical properties [30, 32]. However, the conjugation of graphene with metal NPs is challenging due to both the difficult process of adsorption onto the carbon surface [17] and the uncontrolled size and morphology upon in situ synthesis leading to highly heterogeneous products not suitable for applications requiring mono-disperse material, such as catalytic reactions [5, 18]. We demonstrated that the Prx rings stably adhere onto GO through a "face-on" mode of interaction this is likely due to the larger contact surface involved with respect to an "edge-on" interaction beyond the sticky patches surrounding the ring lumen (Fig. 5).

This oriented interaction allows the GO-bound rings to further interact with other GO in solution resulting in stacking and aggregation. Interestingly, Prx shows a strong activity in reducing GO with striking formation of new C–C bonds in the GO to the detriment of its C–O bonds and concomitant oxidation of the protein's cysteine thiols and methionine thioethers to sulfonates and sulfones, respectively.

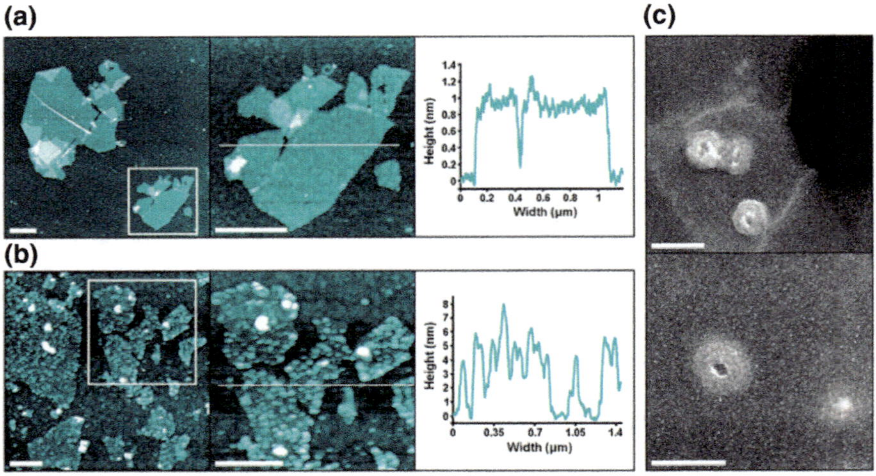

Fig. 5 **a** AFM images of spin-coated GO show single or partially overlapped layers. Zoom details reveal their typical smooth planar surface with height profile (estimated along the *white line*) of ~0.9–1 nm. **b** After dropping *Sm*PrxI, the layered structure of GO is not altered, but its surface becomes highly pitted due to the presence of many protein particles with ~5.1 height and ~20 nm width, according to the dimensions of the *Sm*PrxI ring-like structure. Given the protein coating the height profile of the resulting GO-*Sm*PrxI complexes increases to ~6 nm on average (scale bars = 500 nm). **c** STEM images of individual *Sm*PrxI rings over GO (scale bars = 40 nm). Reproduced from Ardini et al. [4]

This effect enhances the 3D stacks by increasing the hydrophobic π–π interactions between layers [6, 32, 36]. This activity is likely to come from the 50 cysteines and 60 methionines of the Prx ring complex, the latter mostly present at the N-terminal tags of each ring, which are exposed to the GO-Prx interfaces (Fig. 6).

These results are in agreement with the known ability of some sulfur-containing compounds to effectively reduce GO by reducing its epoxide content [7, 8, 10]. These assumptions could be expanded to all proteins characterized by sulfur-containing amino acids. As a matter of fact, Liu et al. found that GO similarly undergoes reduction upon interaction with bovine serum albumin (BSA) [17]. This activity has been ascribed to be at the expense of the 20 tyrosines present in BSA monomer which oxidize to the quinone forms. However, in each BSA monomer (accession code: CAA76847) 35 cysteines and 5 methionines are also present, thus theoretically involved in the reduction of GO likewise we observed for the Prx-GO interaction.

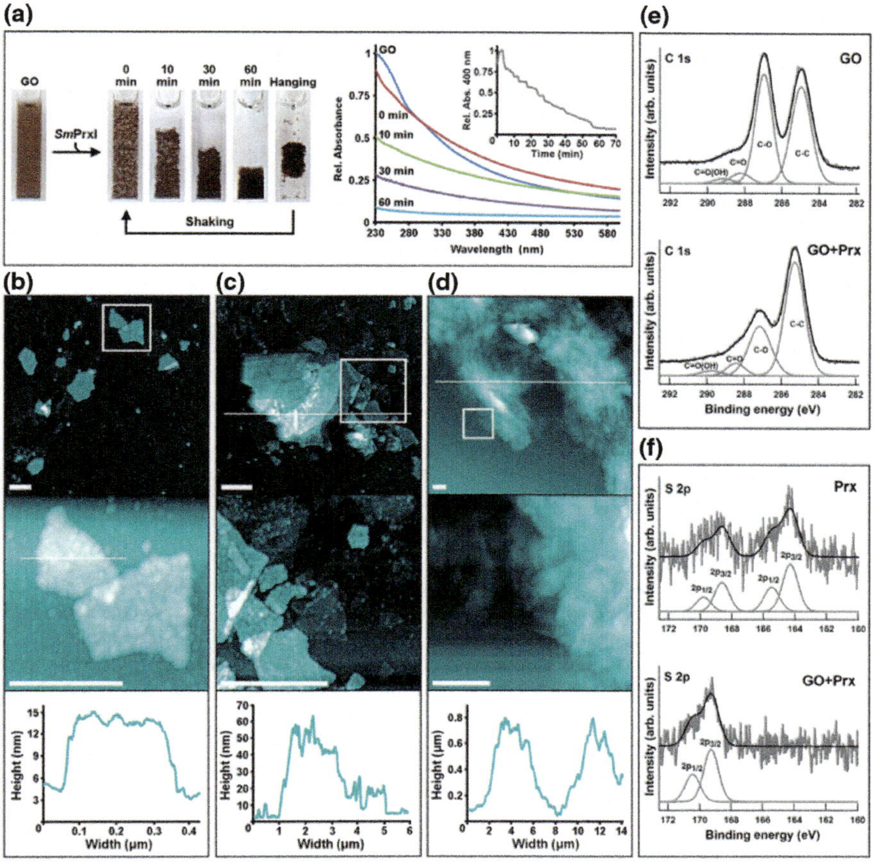

Fig. 6 **a** GO remains soluble in aqueous buffer at pH 7.5. Upon mixing with *Sm*PrxI, it quickly clumps forming visible colloidal particles. Such particles self-assemble within 60 min leading to a free-standing hydrogel-like colloid which is easily broken by shaking though retaining its self-assembly ability to form the aggregate again. The broad optical absorbance of GO is scattered upon mixing with *Sm*PrxI due to the particle formation and progressively decreases as the clumps sink to the bottom of the cuvette forming the colloid. At the end of the process neither GO nor *Sm*PrxI absorbance are significant suggesting they are co-precipitated. The insert illustrates the kinetic trend of the GO absorbance at 400 nm. **b** AFM analysis on the clumped particles shows a heterogeneous population of GO-*Sm*PrxI hybrid composites. The smallest are single or partially overlapped GO layers. Details reveal the presence of a *Sm*PrxI coat making the layer surface highly rough with height ~10 nm pointing to a *Sm*PrxI-GO-*Sm*PrxI sandwich-like architecture (scale bar = 300 nm). **c** Larger composites are found made by stacked multi-layer GO-*Sm*PrxI hybrids where each layer appears covered by several protein rings. Such organization increases the height value to several tens of nanometers (scale bar = 1 μm). **d** Very large complexes with amorphous lobate-like architecture are also observed. Due to their dimensions, both the GO layers and the *Sm*PrxI rings as well as the multi-layer organization are not distinguishable probably because of the low resolution of the scanning tip. These structures show height values of several hundreds of nanometers (scale bar = 1 μm). **e** XPS C 1s core level spectra of GO (*upper spectrum*) and XPS C 1s core level spectra of reduced GO calculated from the C 1s spectra acquired on the GO + *Sm*PrxI sample (*lower spectrum*); the contribution of the C 1s core level signals of SmPrxI were subtracted from both spectra. **f** XPS S 2p core level spectra of Prx (*upper spectrum*) and XPS S 2p core level spectra of GO + *Sm*PrxI (*lower spectrum*). Reproduced from Ardini et al. [4]

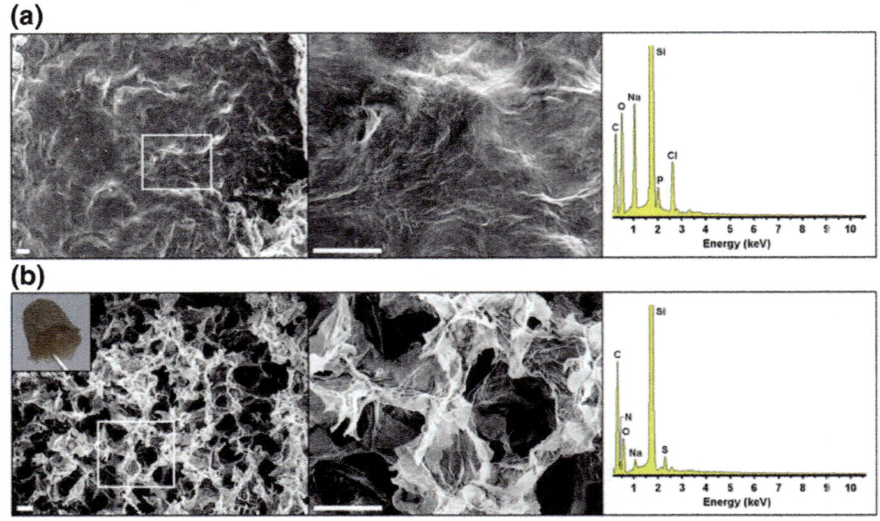

Fig. 7 **a** Freeze-dried GO is imaged as a thin amorphous film under SEM. Zoom details show wrinkles randomly arranged on the carbon surface indicating several overlapped GO layers. The X-ray spectrum from EDS microanalysis proves the carbon and oxygen content in the dried matter at ~0.28 and ~0.52 keV, respectively, besides the substrate and buffer components. **b** The freeze-dried rGO-*Sm*PrxI composite is a soft, compact, free-standing material capable of being manipulated by tweezers. It has an internal microporous architecture with ~5–10 μm cavities whose boundaries are made by multi-layer rGO sheets of several stacked layers. The presence of *Sm*PrxI (not visible without staining) inside the porous network is confirmed by the EDS signals from nitrogen and sulfur at ~0.39 and ~2.3 keV, respectively, which unambiguously belong to the protein (scale bars = 10 μm). Reproduced from Ardini et al. [4]

The architecture of the so-formed Prx-rGO composite clearly shows a microporous network of interconnected multi-layer rGO sheets in which the Prx rings are likely to occupy the interlayer space (Fig. 7).

This system shows a great flexibility in that concerning functionalization with metal NPs For this purposes, the Prx template have been exploited to attach preformed 1.8 nm AuNPs (Fig. 8) or in situ generated PdNPs with roundish shape and dimensions of about 3.3 nm which are synthesized by chemically reducing Pd^{2+} ions linked to both the ring cavity and external surface (Fig. 9). Regardless of either AuNPs or Pd^{2+}/Pd^{0}, the protein retains the sticky behavior towards GO thus allowing a 3D material to be achieved and in both cases it is proven to be

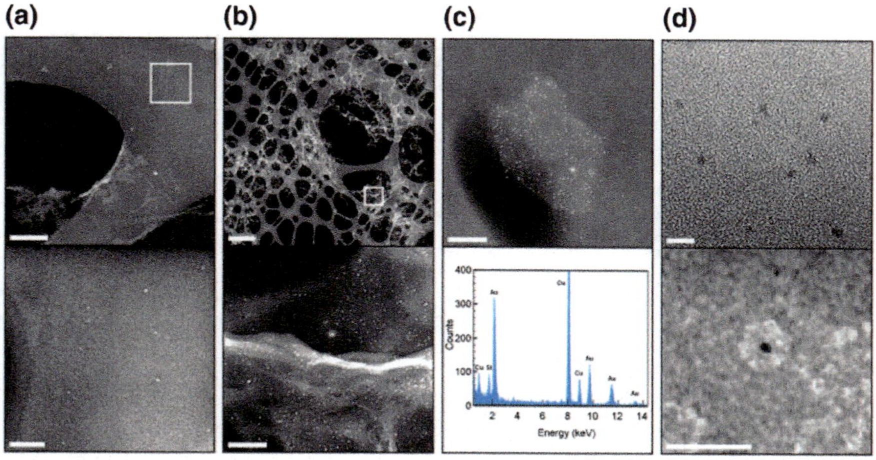

Fig. 8 **a** The AuNPs deposited on rGO layers (scale bars = 100 and 20 nm) are imaged as single shiny spots with around 2 nm diameter evenly and randomly distributed. **b** STEM micrograph showing the rGO-*Sm*PrxI complexes forming multi-layers complexes over the holey carbon film of the TEM grid. Higher magnification images shows the nanoparticle decoration of the rGO flakes (scale bars = 2 μm and 50 nm). **c** STEM image of an individual rGO flake over the amorphous carbon film of the grid, showing selective decoration with AuNPs (scale bar = 50 nm). The presence of gold is proved by EDS analysis performed over the area. **d** HRTEM image of the AuNPs onto the rGO surface in presence of *Sm*PrxI without protein staining (*upper panel*, scale bar = 4 nm). The AuNP within the *Sm*Prxi ring after protein staining is shown (*lower panel*, scale bar = 20 nm). Reproduced from Ardini et al. [4]

mandatory to attach the metal content onto the rGO as there is very poorly direct functionalization of the carbon surface without the protein bridge.

Interestingly, the process of ion adsorption onto GO through Prx moiety is very fast, requiring only a few minutes and this strongly confers to the protein a pivotal role as metal bridge since GO itself requires long-term contact times, higher temperatures and higher Pd^{2+} concentrations even after being modified with metal-binding functionalities [5, 19].

Moreover, characterization of this engineered 3D GO-protein composites will include their potential applications as catalysts, their possible use as biosensors and finally the use of these materials as scaffolds for cellular growth.

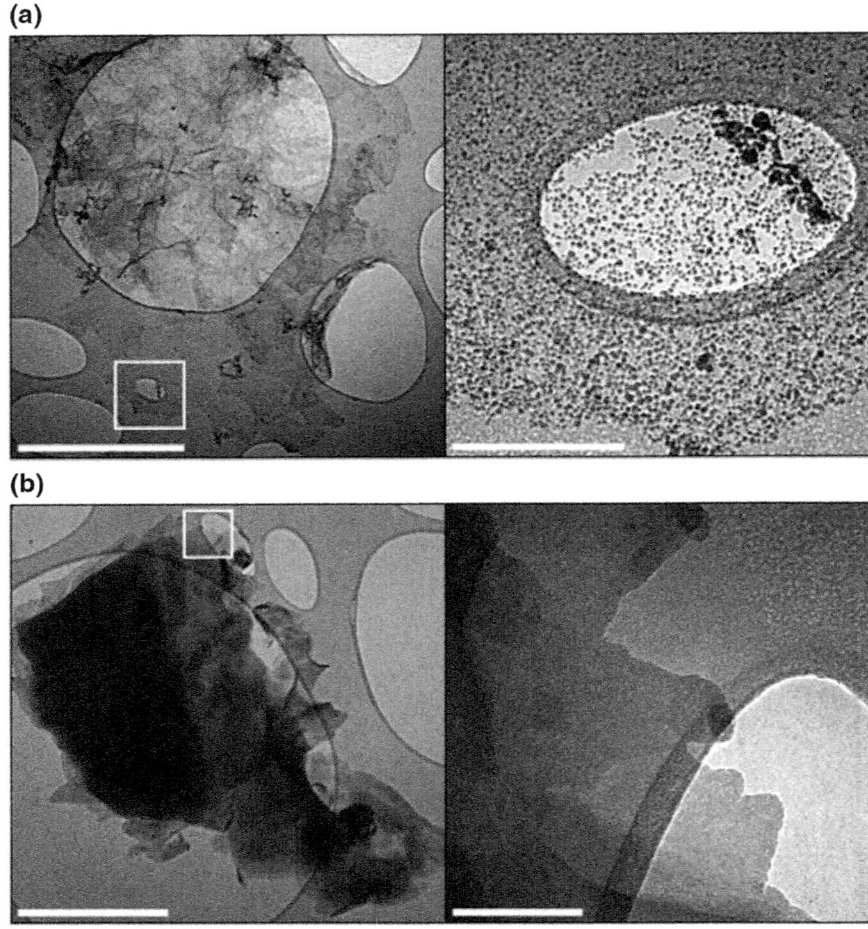

Fig. 9 The PdNPs grow on *Sm*PrxI such that the final composite material shows a very homogeneous arrangement of mono-disperse NPs spreading on the rGO flakes and forming only very few amorphous aggregates (*left panel* scale bar = 1 μm; *right panel* scale bar = 100 nm). **b** TEM micrographs of chemically reduced GO after interaction with Pd^{2+} without *Sm*PrxI. PdNPs are not visible in the 3D GO, demonstrating that the GO alone is not able to effectively retain Pd^{2+} (*left panel* scale bar = 1 μm; *right panel* scale bar = 100 nm). Reproduced from Ardini et al. [4]

References

1. Angelucci, F., Bellelli, A., Ardini, M., Ippoliti, R., Saccoccia, F., Morea, V.: One ring (or two) to hold them all—on the structure and function of protein nanotubes. FEBS J. **282**(15), 2827–2845 (2015)
2. Angelucci, F., Saccoccia, F., Ardini, M., Boumis, G., Brunori, M., Di Leandro, L., Ippoliti, R., Miele, A.E., Natoli, G., Scotti, S., Bellelli, A.: Switching between the alternative structures

and functions of a 2-Cys peroxiredoxin, by site-directed mutagenesis. J. Mol. Biol. **425**(22), 4556–4568 (2013)
3. Ardini, M., Giansanti, F., Di Leandro, L., Pitari, G., Cimini, A., Ottaviano, L., Donarelli, M., Santucci, S., Angelucci, F., Ippoliti, R.: Metal-induced self-assembly of peroxiredoxin as a tool for sorting ultrasmall gold nanoparticles into one-dimensional clusters. Nanoscale **6**, 8052–8061 (2014)
4. Ardini, M., Golia, G., Passaretti, P., Cimini, A., Pitari, G., Giansanti, F., Di Leandro, L., Ottaviano, L., Perrozzi, F., Santucci, S., Morandi, V., Ortolani, L., Christian, M., Treossi, E., Palermo, V., Angelucci, F., Ippoliti, R.: Supramolecular self-assembly of graphene oxide and metal nanoparticles into stacked multilayers by means of a multitasking protein ring. Nanoscale **8**, 6739–6753 (2016)
5. Bai, C., Zhao, Q., Yang, L., Zhang, G., Zhang, F., Fan, X.: Palladium complex immobilized on graphene oxide as an efficient and recyclable catalyst for Suzuki coupling reaction. Catal. Lett. **144**, 1617–1623 (2014)
6. Bai, H., Li, C., Wang, X., Shi, G.: On the gelation of graphene oxide. J. Phys. Chem. C **115**, 5545–5551 (2011)
7. Chen, D., Li, L., Guo, L.: An environment-friendly preparation of reduced graphene oxide nanosheets via amino acid. Nanotechnology **22**, 325601 (2011)
8. Chen, W., Yan, L., Bangal, P.R.: Chemical reduction of graphene oxide to graphene by sulfur-containing compounds. J. Phys. Chem. C **114**, 19885–19890 (2010)
9. de la Rica, R., Matsui, H.: Applications of peptide and protein-based materials in bionanotechnology. Chem. Soc. Rev. **39**, 3499–3509 (2010)
10. Dreyer, D.R., Jia, H., Todd, A.D., Geng, J., Bielawski, C.W.: Graphite oxide: a selective and highly efficient oxidant of thiols and sulfides. Org. Biomol. Chem. **9**, 7292–7295 (2011)
11. Fang, Q., Shen, Y., Chen, B.: Synthesis, decoration and properties of three-dimensional graphene-based macrostructures: a review. Chem. Eng. J. **264**, 753–771 (2015)
12. Feynman, R.P.: There's Plenty of Room at the Bottom. American Physical Society, Pasadena (1959)
13. Gerrard, J.A. (ed.): Protein Nanotechnology: Protocols, Instrumentation, and Applications, Methods in Molecular Biology, vol. 996 (2013)
14. Gradišar, H., Jerala, R.: Self-assembled bionanostructures: proteins following the lead of DNA nanostructures. J. Nanobiotechnol. **12**, 4 (2014)
15. He, D., Marles-Wright, J.: Ferritin family proteins and their use in bionanotechnology. New Biotechnol. **32**, 651–657 (2015)
16. Heddle, J.G.: Protein cages, rings and tubes: useful components of future nanodevices? Nanotechnol. Sci. Appl. 1, 67–78 (2008)
17. Liu, J., Fu, S., Yuan, B., Li, Y., Deng, Z.: Toward a universal "adhesive nanosheet" for the assembly of multiple nanoparticles based on a protein-induced reduction/decoration of graphene oxide. J. Am. Chem. Soc. **132**, 7279–7281 (2010)
18. Lin, J., Mei, T., Lv, M., Zhang, C., Zhao, Z., Wang, X.: Size-controlled PdO/graphene oxides and their reduction products with high catalytic activity. RSC Adv. **4**, 29563–29570 (2014)
19. Liu, L., Liu, S., Zhang, Q., Li, C., Bao, C., Liu, X., Xiao, P.: Adsorption of Au(III), Pd(II), and Pt(IV) from aqueous solution onto graphene oxide. J. Chem. Eng. Data **58**, 209–216 (2013)
20. Ma, Y., Chen, Y.: Three-dimensional graphene networks: synthesis, properties and applications. Natl. Sci. Rev. **2**, 40–53 (2015)
21. Maine, E., Thomas, V.J., Bliemel, M., Murira, A., Utterback, J.: The emergence of the nanobiotechnology industry. Nat. Nanotechnol. **9**, 2–9 (2014)
22. McNeil, S.E.: Nanotechnology for the biologist. J. Leukocyte Biol. **78**, 585–594 (2005)
23. Nardecchia, S., Carriazo, D., Ferrer, M.L., Gutiérrez, M.C., del Monte, F.: Three dimensional macroporous architectures and aerogels built of carbon nanotubes and/or graphene: synthesis and applications. Chem. Soc. Rev. **42**, 794–830 (2013)
24. Palermo, V., Samorì, P.: Molecular self-assembly across multiple length scales. Angew. Chem. Int. Ed. **46**, 4428–4432 (2007)

25. Patil, A.J., Mann, S.: Self-assembly of bio-inorganic nanohybrids using organoclay building blocks. J. Mater. Chem. **18**, 4605–4615 (2008)
26. Patil, A.J., Vickery, J.L., Scott, T.B., Mann, S.: Aqueous stabilization and self-assembly of graphene sheets into layered bio-nanocomposites using DNA. Adv. Mater. **21**, 3159–3164 (2009)
27. Saccoccia, F., Di Micco, P., Boumis, G., Brunori, M., Koutris, I., Miele, A.E., Morea, V., Sriratana, P., Williams, D.L., Bellelli, A., Angelucci, F.: Moonlighting by different stressors: crystal structure of the chaperone species of a 2-Cys peroxiredoxin. Structure. **20**(3), 429–439 (2012)
28. Šimšíková, M.: Interaction of graphene oxide with albumins: effect of size, pH, and temperature. Arch. Biochem. Biophys. **593**, 69–79 (2016)
29. Takeda, Y., Mae, S., Kajikawa, Y., Matsushima, K.: Nanobiotechnology as an emerging research domain from nanotechnology: a bibliometric approach. Scientometrics **80**, 25–40 (2009)
30. Wang, Z., Lv, X., Chen, Y., Liu, D., Xu, X.G., Palmore, T.R., Hurt, R.H.: Crumpled graphene nanoreactors. Nanoscale **7**, 10267–10278 (2015)
31. Wei, X.Q., Hao, L.Y., Shao, X.R., Zhang, Q., Jia, X.Q., Zhang, Z.R., Lin, Y.F., Peng, Q.: Insight into the interaction of graphene oxide with serum proteins and the impact of the degree of reduction and concentration. ACS Appl. Mater. Interfaces. **7**, 13367–13374 (2015)
32. Whitby, R.L.D.: Chemical control of graphene architecture: tailoring shape and properties. ACS Nano **8**, 9733–9754 (2014)
33. Whitesides, G.M., Boncheva, M.: Beyond molecules: self-assembly of mesoscopic and macroscopic components. PNAS **99**, 4769–4774 (2002)
34. Whitesides, G.M., Mathias, J.P., Seto, C.T.: Molecular self-assembly and nanochemistry: a chemical strategy for the synthesis of nanostructures. Science **254**, 1312–1319 (1991)
35. Xiao, Z., Wedd, A.G.: The challenges of determining metal-protein affinities. Nat. Prod. Rep. **27**, 768–789 (2010)
36. Xu, Y., Wu, Q., Sun, Y., Bai, H., Shi, G.: Three-dimensional self-assembly of graphene oxide and DNA into multifunctional hydrogels. ACS Nano **4**, 7358–7362 (2010)
37. Zeng, Q., Cheng, J., Tang, L., Liu, X., Liu, Y., Li, J., Jiang, J.: Self-assembled graphene-enzyme hierarchical nanostructures for electrochemical biosensing. Adv. Funct. Mater. **20**, 3366–3372 (2010)
38. Zhang, Y., Wu, C., Guo, S., Zhang, J.: Interactions of graphene and graphene oxide with proteins and peptides. Nanotechnol. Rev. **2**, 27–45 (2013)
39. Zhao, Z., Wang, X., Qiu, J., Lin, J., Xu, D., Zhang, C., Lv, M., Yang, X.: Three-dimensional graphene-based hydrogel/aerogel materials. Rev. Adv. Mater. Sci. **36**, 137–151 (2014)

Graphene- and Carbon Nanotubes-Yeast Bionicomposites

L. Valentini, S. Bittolo Bon, S. Signetti and N.M. Pugno

Abstract Nature offers us an enormous amount of ready-to-use templates with various morphologies and functionalities, which can be successfully utilized in fabrication of biosensors, tissue engineering, and microelectronics. The directed combination of such natural templates with graphene or carbon nanotubes results in the development of a novel material which uses the features of both. We produced hybrid materials by giving to microorganisms the nutrient to grow together with graphene nanoplatelets and carbon nanotubes. Such hybrid materials can be considered as bionic because they have the benefits of both biological world which can self-organize and that of non-living materials, which couple functions such as self-healing and electronic transport.

1 Introduction

Biogenic materials are often formed at the nanometer scale through diverse metabolic activities and by passive surface reactions on cell walls or extracellular structures [1, 2]. The exploitation of the processes used by microorganisms to digest

L. Valentini (✉) · S. Bittolo Bon
Dipartimento di Ingegneria Civile e Ambientale, Università di Perugia,
UdR INSTM di Perugia, strada di Pentima 4, 05100 Terni, Italy
e-mail: luca.valentini@unipg.it

S. Bittolo Bon · S. Signetti · N.M. Pugno (✉)
Laboratory of Bio-Inspired and Graphene Nanomechanics, Department of Civil,
Environmental and Mechanical Engineering, University of Trento, via Mesiano 77,
38123 Trento, Italy
e-mail: nicola.pugno@unitn.it

N.M. Pugno
School of Engineering and Materials Science, Queen Mary University of London,
Mile End Road, London E1 4NS, UK

N.M. Pugno
Ket-Lab, Edoardo Amaldi Foundation, Italian Space Agency, via del Politecnico snc, 00133
Rome, Italy

V. Morandi and L. Ottaviano (eds.), *GraphITA*, Carbon Nanostructures,
DOI 10.1007/978-3-319-58134-7_16

nutrients for their growth can be a viable method for the formation of a wide range of so-called biogenic materials that have unique mechanical and physical properties that are not produced by abiotic processes [3]. The interaction between living organisms and inorganic materials, including inorganic nanomaterials with higher electrical conductivity, could facilitate extracellular electron transfer as well as the improving of the mechanical properties [4, 5].

Baker's yeast has been used for centuries in wine and bread making and it consists of *Saccharomyces cerevisiae (SaC)*, a unicellular microorganism [6–8]. Yeast is a cellular factory and reproduces asexually through a process called "budding," in which a "mother" cell grows a "daughter" cell that separates to become fully independent. We exploited yeast fermentation to couple graphene on its cells' walls. The technique uses the commercially available and inexpensive bread yeast and can be applied to any living cells without special surface modification of graphene. In turn, this process could lead to the development of hierarchical and interactive structures programmed to self-assemble into specific patterns, such as those on strain sensors, and of self-healing materials capable to sense and repair mechanical damage.

The scientific core is to address the synthesis of hybrid living materials by means of "cold fermentation" processes through alcoholic fermentation and large-scale living microbes' factories. In this context, this research could have future implications of bio-materials in the manufacture of physical smart and multifunctional objects. Encapsulation of the cell with an inorganic shell has already been found to enhance the robustness of the cell and improve its endurance in harsh environments [9]. Some materials such as silica and calcium phosphate have already been used as encapsulating materials to improve the cell stability [10, 11]. Our works [4, 5, 12, 13] present the first attempts in producing bionicomposites and [4, 5] are revised in this book chapter.

2 Graphene-Yeast Bionicomposites

For the preparation of the composite film, 40 mg of graphene nanoplatelets (GNPs, purchased from Cheaptubes®, thickness 8–15 nm) were dispersed for 3 h at room temperature in 40 mL of water using a sonication bath. *SaC* based commercial beer yeast extract was used as the medium for fermentation. The dispersion of GNPs (1 mg/ml) was then added to the yeast solution and stirred at 110 rpm at 30 °C for 1 h. To start the fermentation, sugar (i.e., sucrose, 4 g) was added to the yeast/GNPs solution. The dispersion was heated at 35 °C to allow the fermentation process. The solid films were obtained by evaporating the water in excess, leaving the solutions in a sterilized silicon rubber mold at 30 °C in inert atmosphere for two nights. Field emission scanning microscopy (FESEM) was used to investigate the cross section of the samples obtained by fracture in liquid nitrogen. The current–voltage characteristic was performed by a computer controlled Keithley 4200 Source Measure Unit. The electrical conductivity of the samples was monitored, at room temperature, by applying a sweeping DC electric voltage across the sample. The tensile properties of films, i.e., Young's modulus and tensile strength, were then measured using a universal tensile testing machine (Lloyd Instr. LR30 K) with a 50 N static load cell. The film samples were cut into strips (30 mm × 12 mm). The gauge length was 20 mm, and the extension rate was set at 2 mm/min.

The shape of the yeast cells was found to be spheroidal (Fig. 1a). According to previous papers, *SaC* multiplies with a process where a daughter cell is initiated as growth from the mother cell (Fig. 1b, c) [14]. The round protrusion on the cell surface, visible in Fig. 1b, is a bud scar. The bud scar forms on the cell after the process of division has taken place.

The effect on the film morphology when GNPs were added during fermentation is reported in Fig. 1c. The FESEM analysis on the GNP/yeast cell surface indicates the formation of wrinkles which arise from compressive stress when a soft-matter core (i.e., *SaC* yeast cell in our case) is coupled with a thin, high modulus sheet (i.e., GNPs) [15]. The physics and geometry of such system indicates that the wavelength λ of the wrinkles can be estimated as follows:

$$\lambda = 2\pi t \left[\frac{(1 - \upsilon_S^2)}{(1 - \upsilon^2)} \frac{E}{3E_S} \right]^{1/3} \tag{1}$$

where t is the thickness of the GNPs film, E and ν are respectively Young's modulus and Poisson's ratio of the film, while E_s and ν_s refer to the substrate counterparts. Assuming an average thickness t = 11.6 nm, E = 50 GPa, ν = 0.165, E_s = 112 MPa, and ν_s = 0.5 [16] a wrinkling wavelength $\lambda \approx 38$ nm is obtained, in agreement with our experimental value of 30–70 nm (Fig. 1d).

Cell wall can be considered as an elastic encapsulating layer allowing the exchange of substance with the surroundings [16]. *SaC* is a type of yeast widely used in the conversion of sugar because of its ability to produce high amounts of ethanol and glucose as by products [17]. Thus the production of glucose during *SaC* fermentation can be used to couple graphene nanoplatelets onto the yeast cell wall and, when sugar is exhausted, to obtain bionic composite films.

The measured composite, fracture strength σ_f and Young's modulus E, can be related to the properties of the constituent phases via rules of mixture:

$$\sigma_f = f \sigma_{f,\mathrm{GNPs}} + (1 - f)\sigma_{f,\mathrm{yeast}} \tag{2a}$$

$$E = f E_{\mathrm{GNPs}} + (1 - f)E_{\mathrm{yeast}} \tag{2b}$$

where f is the volumetric fraction of the GNP film and the subscripts refer to each component of the bionic composite. Parallel, a simple system of two fermented cells was modeled via Finite Element Method (FEM). The shape of the cells within the ensemble is approximated as hexagonal prism (Fig. 2a). The dimensions of the cell, mother and daughter, were taken from the work of Ahmad et al. [16] which are

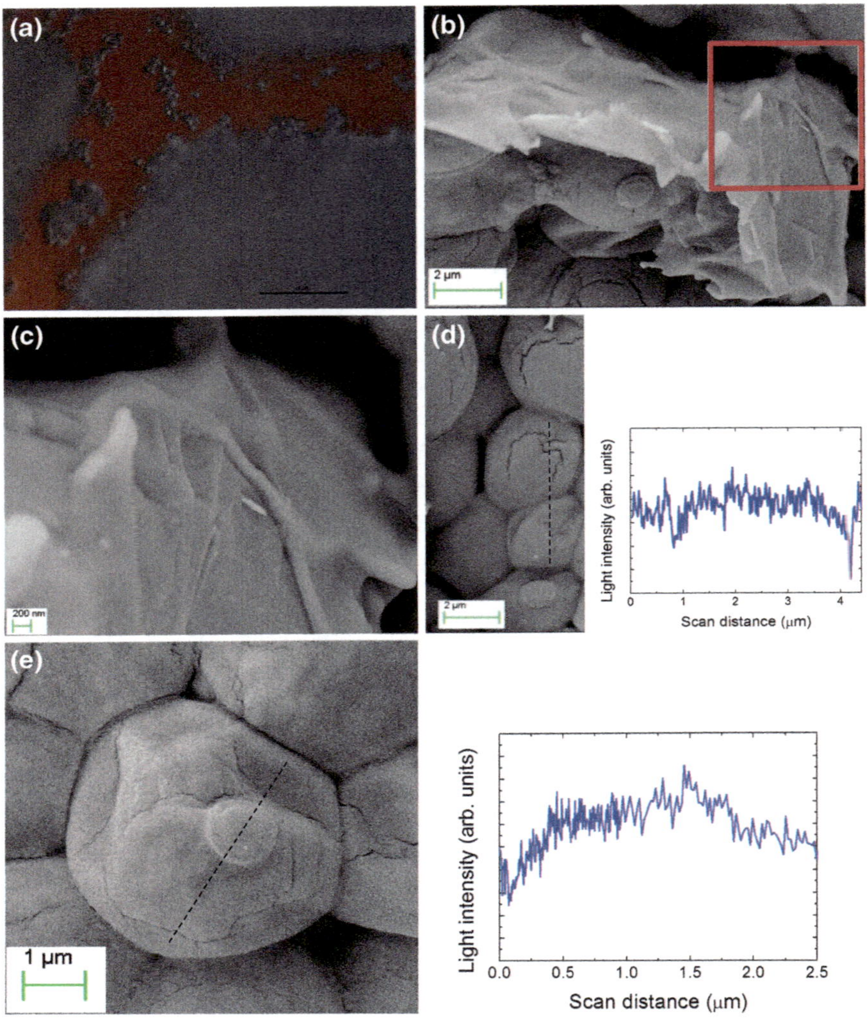

Fig. 1 **a** Optical image of yeast cells (the scale bar indicates 60 µm). **b** FESEM image of the GNPs/yeast cell sample and **c** its magnified view. **d** FESEM image and line scan of the surface of the damaged yeast/GNPs sample. **e** FESEM image and line scan of the surface of the repaired yeast/GNPs sample [5]

comparable to our samples [4]. As regards the yeast constitutive behavior, the curve obtained from the pristine yeast [4] has been adopted as input for simulations, using an elastic material with finite strain capability. The cells are modeled with solid elements. The load is applied as imposed displacements on the lateral face of the cells (Fig. 2a). The interface between the two cells was modeled via a cohesive

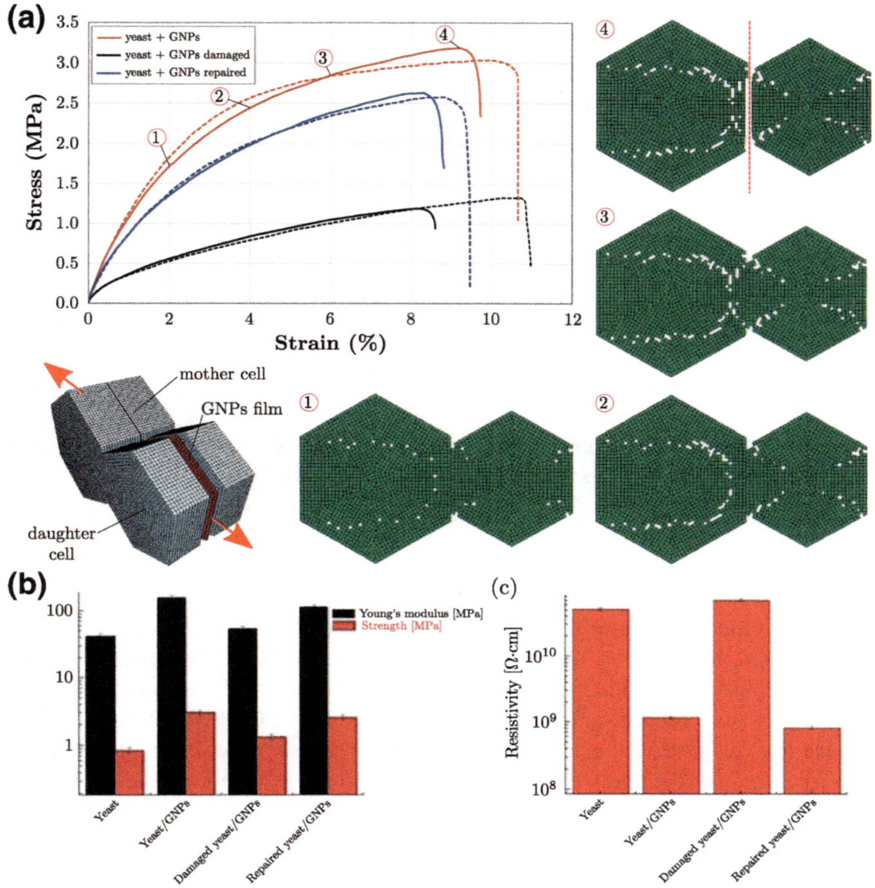

Fig. 2 **a** Simulation of the base "sandwich" composite element made up of a mother cell and a daughter cell with a GNP film in between of 11.5 nm thickness. The simulation-derived curves (continuous) are compared with the corresponding experimental counterpart (*dashed*). The evolution of the fracture pattern in the graphene layers at four different states for increasing tension is depicted: composite failure occurs because of film rupture (state 4) along the interface between the two cells. **b** Mechanical characteristic and **c** electrical resistivity values of neat yeast, yeast/GNPs, damaged and repaired yeast/GNPs samples, respectively [5]

zone model (CZM) based contact for which the force-separation curve was derived in [5] by the authors and associated with an estimated fracture energy of $G_{\text{f-yeast}} = 0.0193$ N/m. GNPs film has been modeled with shell elements assuming an average thickness of 11.6 nm. This corresponds to a total of 34 layers of graphene. Two half cells with a GNPs film in between has been modeled in a sort of "sandwich" structure with periodic boundary conditions at the symmetry planes:

by doing this we simulate an infinite multilayer alternation of yeast cell layer and GNPs film. Constitutive behavior of graphene has been taken from Xu et al. [18], operating a scaling on the failure strength of the curves there reported in order to minimize the deviation of simulation curves with respect to the experimental ones and thus have an indirect estimate of the mechanical properties of GNPs. Perfect bonding was assumed at the interface between the GNPs film and the cell wall, being the constitutive behavior of this interface unknown.

Figure 2a shows the comparison between the curves derived from experiments and simulated tensile tests. The mechanical properties determined from simulation are well comparable with the one estimated with the rule of mixture (Table 1), confirming the mechanical interaction between the GNPs film and yeast cells. A first appearance of diagonal cracks within graphene is followed by their coalescence in the correspondence of the interface between the two cells at which global failure of the composite occurs. This is qualitative comparable with the FESEM experimental images (Fig. 1d, e) and, as obtained for yeast-CNTs composites [4], global failure is governed by interface rather than cell rupture.

Under high-vacuum conditions of FESEM, the formation of cracks on the cell surface indicates a hypo-osmotic shock that is due to the coupling between the GNPs sheets and the cell wall hindering cell dehydration through its membrane (Fig. 1d). We observe that such crack formation can be healed by placing the damaged composite film in aqueous solution of sucrose and GNPs for few minutes; as suggested by FESEM analysis (Fig. 1e), the cracks almost disappear. This self-healing effect is evident on the mechanical and electrical characteristics of the graphene-bionic composite, as reported in Fig. 2b, c. The GNP sheets cover the cell surface and this gives the electrical conductivity and mechanical strength across the cell surface; the decrease of the electrical conductivity and of the mechanical strength as well is attributed to the loss of the graphene connecting network. The nutrient broth reconstructs the GNPs shell and the strain-induced effect on GNP sheets is canceled, partially restoring the mechanical and electrical properties of the undamaged composite.

Table 1 GNPs properties determined from experiments through rules of mixture or from numerical simulations. For both estimates of GNPs properties we assumed an average film thickness $t_{GNP} = 11.5$ nm

Material	Experiments				Simulations			
	σ_f (MPa)	E (MPa)	$\sigma_{f\text{-}GNP}$ (MPa)	E_{GNP} (GPa)	σ_f (MPa)	E (MPa)	$\sigma_{f\text{-}GNP}$ (MPa)	E_{GNP} (GPa)
Yeast+GNPs	3.1	159	909	48.8	3.2	135	882	50.1
Yeast+GNPs damaged	1.3	54	196	5.1	1.2	50	126	5.3
Yeast+GNP repaired	2.6	115	719	30.5	2.6	107	670	31.7

3 CNTs-Yeast Bionicomposites

An analogous procedure was used to produce CNTs-based yeast bionicomposites (Fig. 3a) [4]. Water dispersion of CNTs (1 mg/ml) were prepared by tip sonication. This dispersion of CNTs was then added to the yeast solution and stirred at 110 rpm at 30 °C for 1 h. Then the yeast/CNTs solution was put into a sterilized circular aluminum mold, and the liquid medium was left to evaporate at 30 °C during the night. In another sterilized flask the same procedure was adopted by adding sugar to promote the fermentation of the yeast/CNTs solution. After fermentation, the liquid medium was dried in a sterilized mold and a composite film was obtained. This procedure results in a complete encapsulation of CNTs by the yeast cells (Fig. 3c, d).

The strength of the composite can be estimated according to Eq. 2a in which the strength of the GNPs is substituted by the pull-out strength of CNTs, derived from the pull-out energy G at the CNT-yeast interface. The fiber volume fraction was estimated to be $f \approx 0.021$ from FESEM images (encircled region in Fig. 3b). The strength of the pristine yeast and for CNTs-yeast fermented composite were respectively 0.0297 MPa and 0.240 MPa obtaining a value of CNTs pull-out strength $\sigma_{f,CNTpo} \approx 10.3$ MPa. From this calculation it is clear how the strength of the yeast is much lower than the pull-out tension, indicating that the strength of the composite is governed by the adhesion energy between the CNTs and the matrix.

Analogously to the previous case, a FEM model of a system of two cells was developed. The interface between the two cells implements a cohesive zone model based contact and its adhesion energy, $G_{f,comp}$ assumed of pure Mode I fracture, is a combination by a rule of mixture of the interface fracture energy of the pristine

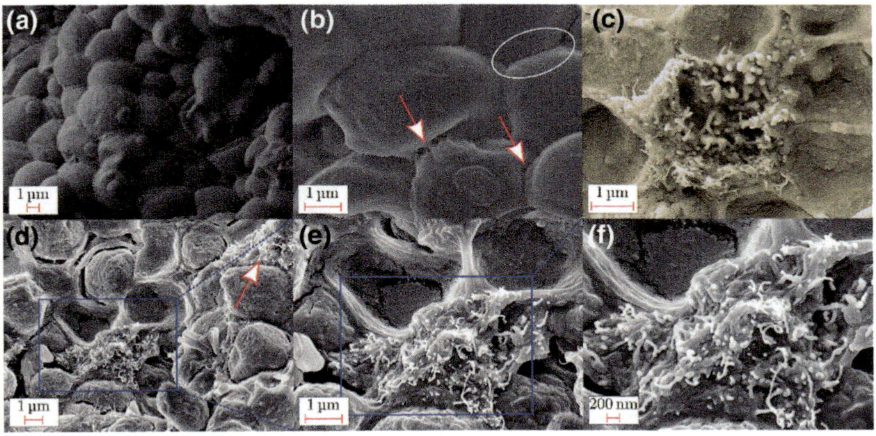

Fig. 3 **a** FESEM image of Saccharomyces cerevisiae. **b** FESEM image showing CNTs bridging yeast cells. The arrows indicate the CNTs bridging the yeast cells. **c** FESEM image of the cross section of the fermented yeast/CNTs film after prolonged exposure to FESEM where CNTs protruding from a broken yeast cell are visible. **d, f** FESEM images of the yeast/CNTs film without fermentation taken at different magnifications [4]

yeast interface (Fig. 4a) and of the adhesion energy of CNTs determined above with the pull-out model (details can be found in Ref. [4]). Under tensile load, failure occurs when for the current interface energy G is $G/G_{\mathrm{f,comp}} > 1$. Figure 4 shows the superposition of the FEM simulations curves on experimental ones. Figure 4a refers to mother–daughter cells in fermented yeast, while in Fig. 4b the cases of equal

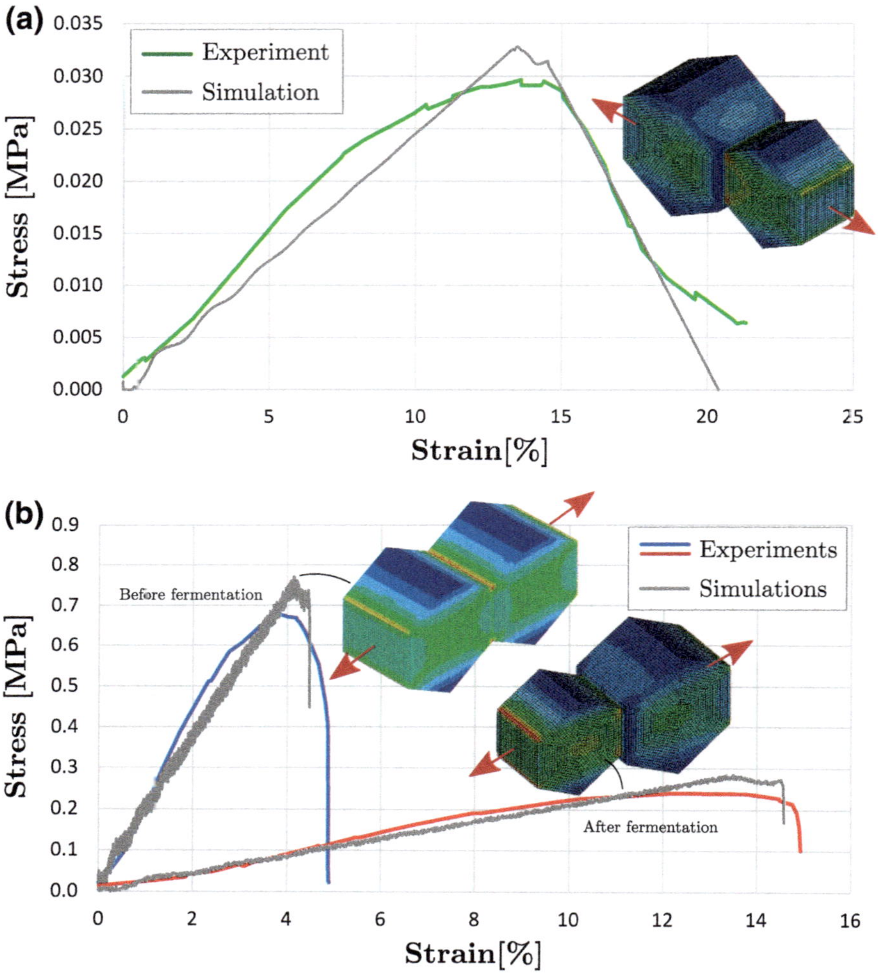

Fig. 4 Stress-strain curves obtained from tensile tests on **a** fermented yeast sample (*green curve*) and **b** yeast/CNT composites prepared before fermentation (*blue curve*) and after fermentation (*red curve*), respectively. Note that for the fermented system two different cell type in terms of dimension and mechanical properties were adopted in order to take into account the mother-daughter cells systems. The comparisons with the result of FEM simulation of traction of a two-cell system are also reported [4]

cells with CNTs before fermentation and mother–daughter cells with CNTs after fermentation are reported. A decrease in the CNTs volume fraction after fermentation is estimated. This result is interpreted with a higher contact area between cells due to the increase of the cells volume after fermentation explaining the lower failure stress and the higher failure strain observed for the composite film after fermentation. Further investigations on this are required.

Current–voltage characteristic of the yeast-CNTs composite was then measured (Fig. 5). We observed that the increase in conductivity is more prominent with the CNTs. Furthermore, the addition of sucrose to the neat yeast resulted in a better conductivity being this effect more pronounced for the fermented yeast/CNTs system. The source of this behavior is the bridging of conductive CNTs within the yeast during the fermentation that improves the percolation pattern for electron transfer. On the contrary the formation and confinement of isolated CNTs bundles between the cell walls (Fig. 3d, f) result in an interruption of the percolation pattern leading to a lower conductivity for the non-fermented yeast/CNTs sample.

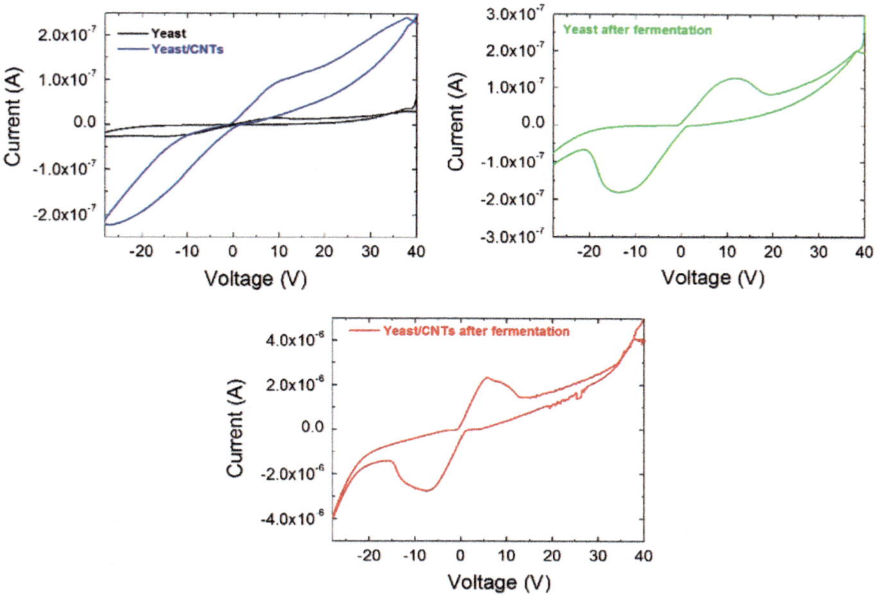

Fig. 5 I-V characteristics of the prepared samples [4]

4 Conclusions

In this study, we demonstrated that the fermentation of commercially available and inexpensive bread yeast with graphene sheets leads to the development of interactive bionic structures programmed to self-assemble into living templates. Encapsulation of the cell with graphene nanoplatelets has been found to enhance the yeast mechanical strength as well as the electrical conductivity. The main result of this study is the development of nanostructured shells that do not disturb the cell proliferation which acts as an electrical conductor and as mechanical reinforcement. At the same time they show self-healing capacity, being able to re-assemble onto the cell surface when the encapsulation is broken by external harsh environment. Similar results were also obtained in CNTs-based fermented yeast composites, in which the conductivity and mechanical properties are enhanced by a nanotube bridging mechanism. In this context, these works [5, 6] aim at the development of future applications of such bionic materials, e.g., in the manufacture of physical smart objects with multifunctional properties (bioelectronics), in the fields of self-healing and functional bionic composites.

Acknowledgements NMP is supported by the European Research Council (ERC PoC 2015 SILKENE nr. 693670) and by the European Commission H2020 under the Graphene Flagship (WP14 "Polymer composites", n. 696656) and under the FET Proactive ("Neurofibres" n.o. 732344). SS acknowledges support from SILKENE.

References

1. Newman, D.K.: Microbiology—How bacteria respire minerals. Science **292**, 1312 (2001)
2. Lowenstam, H.A.: Minerals formed by organisms. Science **211**, 1126 (1981)
3. Fortin, D.: What biogenic minerals tell us. Science **303**, 1618 (2004)
4. Valentini, L., Bittolo Bon, S., Signetti, S., Tripathi, M., Iacob, E., Pugno, N.M.: Fermentation based carbon nanotube multifunctional bionic composites. Sci. Rep. **6**, 27031 (2016)
5. Valentini, L., Bittolo Bon, S., Signetti, S., Pugno, N.M.: Graphene-based bionic composites with multifunctional and repairing properties. ACS Appl. Mater. Interf. **8**, 7607 (2016)
6. Chen, Y., Daviet, L., Schalk, M., Siewers, V., Nielsen, J.: Establishing a platform cell factory through engineering of yeast acetyl-CoA metabolism. Metab. Eng. **15**, 48 (2013)
7. Hong, K.K., Nielsen, J.: Metabolic engineering of Saccharomyces cerevisiae: a key cell factory platform for future biorefineries. Cell. Mol. Life Sci. **69**, 2671 (2012)
8. Runguphan, W., Keasling, J.D.: Metabolic engineering of Saccharomyces cerevisiae for production of fatty acid-derived biofuels and chemicals. Metab. Eng. **21**, 103 (2014)
9. Kempaiah, R., Chung, A., Maheshwari, V.: Graphene as cellular interface: electromechanical coupling with cells. ACS Nano **5**, 6025 (2011)
10. Yang, S.H., Ko, E.H., Jung, Y.H., Choi, I.S.: Bioinspired functionalization of silica-encapsulated yeast cells. Angew. Chem. Int. Ed. **50**, 6115 (2011)
11. Wang, B., Liu, P., Jiang, W.G., Pan, H.H., Xu, X.R., Tang, R.K.: Yeast cells with an artificial mineral shell: protection and modification of living cells by biomimetic mineralization. Angew. Chem. Int. Ed. **47**, 3560 (2008)

12. Lepore, E., Bonaccorso, F., Bruna, M., Bosia, F., Taioli, S., Garberoglio, G., Ferrari, A.C., Pugno, N.M.: Silk reinforced with graphene or carbon nanotubes spun by spiders. arXiv:1504.06751 (2015)
13. Valentini, L., Bittolo Bon, S., Pugno, N.M.: Graphene and carbon nanotube auxetic rubber bionic composites with negative variation of the electrical resistance and comparison with their nonbionic counterparts. Adv. Funct. Mater., 1606526 (2017)
14. Broach, J.R., Jones, E.W., Pringle, J.R.: The Molecular and Cellular Biology of the Yeast Saccharomyces, Vol. 1. Genome Dynamics, Protein Synthesis, and Energetics. Cold Spring Harbor, NY, Cold Spring Harbor Laboratory Press (1991)
15. Cerda, E., Mahadevan, L.: Geometry and physics of wrinkling. Phys. Rev. Lett. **90**, 074302 (2003)
16. Ahmad, M.R., Nakajima, M., Kojima, S., Homma, M., Fukuda, T.: In situ single cell mechanics characterization of yeast cells using nanoneedles inside environmental SEM. IEEE T. Nanotechnol. **7**, 607 (2008)
17. Balat, M., Balat, H., Oz, C.: Progress in bioethanol processing. Prog. Energy Combust. Sci. **34**, 551 (2008)
18. Xu, M., Paci, T.J., Oswald, J., Belytschko, T.: A constitutive equation for graphene based on density functional theory. Int. J. Solids Struct. **49**, 2582 (2012)

Printed by Printforce, the Netherlands